weather

Ralph Hardy

TEACH YOURSELF BOOKS

For Elizabeth

Long-renowned as the authoritative source for self-guided learning – with more than 30 million copies sold worldwide – the *Teach Yourself* series includes over 200 titles in the fields of languages, crafts, hobbies, sports, and other leisure activities.

Library of Congress Catalog Card Number: 96-68485

First published in UK 1996 by Hodder Headline Plc, 338 Euston Road, London NW1 3BH

First published in US 1996 by NTC Publishing Group
An imprint of NTC/Contemporary Publishing Company
4255 West Touhy Avenue, Lincolnwood (Chicago), Illinois 60646-1975 U.S.A.

Typeset by Transet Limited, Coventry, England.
Printed in Great Britain by Cox & Wyman Ltd, Reading, Berkshire.

Impression number 10 9 8 7 6 5 4
Year   2002 2001 2000 1999 1998 1997

CONTENTS

INTRODUCTION

The aim of this book is to provide a comprehensive and yet easily understood account of weather phenomena and climate round the world, together with the reasons behind it all. Modern as well as traditional methods of observing and forecasting are discussed, but the emphasis is on understanding what we see around us and how it fits into the global picture. Since the 1940s increased knowledge of the atmosphere has enabled sophisticated mathematical models to be devised. These, combined with increasingly powerful computers, have dramatically advanced weather forecasting. Readers will find that an understanding of how the atmosphere works will enable weather forecasts to be interpreted more easily, in respect of their own locality.

Without weather, life on Earth as we know it could not exist. It would be a barren, inhospitable desert, unbearably hot in low latitudes and impossibly cold from middle latitudes polewards.

Our world and its atmosphere can be regarded as an engine. Its fuel is power from the sun, and its output is the life of every living creature and plant. The moving parts of this engine lie within the restless atmosphere – a relatively thin skin of air, in which wind and weather systems ceaselessly spread excess heat from the tropical regions towards the poles. The beautiful cloud patterns we see on satellite pictures are components of large-scale weather systems viewed from space. Within them, and in the quieter areas between, lie smaller-scale but no less interesting or important phenomena, all components of our magnificent weather machine.

Weather has an important place in the lives of us all. It affects agriculture, transport, commerce and industry. It influences our housing, dress, food, leisure and holidays! We cannot tame it, but we can observe and measure it, forecast and use it, and where necessary take precautions against it. We also need to take a longer view; to be aware of environmental changes which may influence future weather and climate. These vary from the local effects of air pollution and acid rain, to global warming and climate change, and also the possible decrease of the protective ozone layer in the upper atmosphere.

No matter how complicated a machine may be, it is made up of many simple components each precisely engineered. The whole may be understood from a knowledge of these components and how they fit together. This applies equally to weather and the general circulation of the atmosphere. No knowledge of advanced mathematics or physics is required, just the ability to see how each small cog meshes with others to drive our restless atmosphere.

Teach Yourself Weather is presented in two parts:

Part One discusses the driving forces behind our weather and major weather systems. It is intended to be read sequentially. Within it weather and climate are explained from the largest-scale influences – the Earth's orbit around the sun – down to the smallest component – the remarkable water molecule. The pieces are examined and assembled to explain the various weather systems and their characteristics.

Part Two deals with many weather related topics not covered in Part One. In this section each chapter stands alone to provide a source of easily accessed information focused on a certain aspect of weather. Part Two is intended to be dipped into as and when required and not necessarily read sequentially. Where more general guidance may be needed the reader is referred to the appropriate chapter in Part One, to avoid duplication.

The Index contains many of the most common meteorological terms and jargon. Each entry gives the page where that term is defined or explained. That page only will have the term emboldened to aid reference. This substitutes for a glossary with the advantage of keeping explanations in context and avoiding repetition.

To prevent making the text unwieldy, discussion usually focuses on the Northern hemisphere except where stated otherwise. The usual meteorological conventions are adopted; in particular winds are always described in terms of the direction from which they blow; hence a cold northerly wind in the northern hemisphere is the exact equivalent of a cold southerly in the southern hemisphere, both blowing from the direction of the pole towards the equator.

Units used are usually in the centimetre/gram/second (cgs) system except where common meteorological practice dictates otherwise, in which case they are defined when introduced.

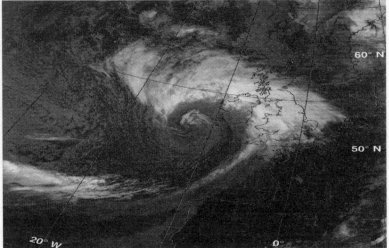

Simultaneous satellite pictures in visible (bottom) and infrared (top) wavebands, taken on 1 August 1986 at 14:55 UTC. These show extensive cloud, associated with a deep depression near south-west Ireland, across most of the British Isles, and discrete shower clouds over the Atlantic south and west of the centre. As explained in Chapter 6, warmer low cloud shows in the infrared as a darker grey, whereas in the visible waveband all cloud is white. This contrast can be seen where cloud spirals in towards the centre of the low.

Part One

THE ATMOSPHERE AND MAJOR WEATHER SYSTEMS

1

THE EARTH, THE SUN AND THE SEASONS

Earth, the world upon which we live, is about 4400 million years old, and has changed greatly over that time. Fortunately changes have historically been extremely slow, and for our purposes it can be regarded as a fixed if not stationary frame of reference. Earth is not exactly spherical, its circumference being 40 076 km (24 903 miles) around the equator and 40 008 km (24 860 miles) around the poles, but again the difference is negligible for most purposes. The largest ocean, the Pacific, is greater in area than all the continents put together; and overall more than 70 per cent of the Earth's surface is water.

All the planets in the solar system, including our Earth, orbit the sun. The Earth's orbit takes one year: that is 365 days, 5 hours, 48 minutes and 46 seconds approximately! If we imagine the Earth to be orbiting around the circumference of a flat plate, with the sun at its centre, this plate is called the **ecliptic**. The plate would need to be slightly elliptical because the Earth's orbit is not quite a circle. We pass closest to the sun during the northern hemisphere winter at a distance of 146.4 million km (90.9 million miles), and furthest away from the sun in the southern hemisphere winter at about 151.2 million km (93.9 million miles). This relatively small difference has little effect on our climate, but it would if it was much larger or smaller. Indeed life as we know it is impossible on any other planet in our solar system, because they are closer to the sun and impossibly hot, or further away and unimaginably cold. Furthermore none has our life-support system of the atmosphere and weather within it.

Our Earth is unique. It is sufficiently near to the sun to ensure that it does not entirely freeze, and far enough away for it not to boil. Furthermore it has a force of gravity which is sufficient to stop our life-giving atmosphere drifting off into space, but not so great that it prevents us from moving around. None of this may seem remarkable, but we must remember that our relatively small planet orbiting a modest star is but one of countless billions in our galaxy alone, and that beyond the Milky Way there are billions of other galaxies. Since no evidence has yet been found of life anywhere else in the universe, it appears that our circumstances are rare indeed!

Virtually all the heat at the Earth's surface and throughout the atmosphere comes from the sun. Contributions from the Earth's hot, molten core, can occasionally be seen bursting from active volcanoes and in some countries from hot water geysers originating deep underground. Nevertheless these sources are negligible in comparison with heat from the sun.

The Earth not only orbits the sun, but also rotates once each day about its own axis, which runs from pole to pole. We conventionally think of the North Pole at the top and the South Pole at the bottom of the globe, but the axis is not perpendicular to the ecliptic – it tilts at an angle of 23.5 degrees. For half of the year (summer in the northern hemisphere and winter in the south) the North Pole tilts towards the sun. In the other six months the South Pole is towards the sun, and the seasons are reversed (see Figure 1.1).

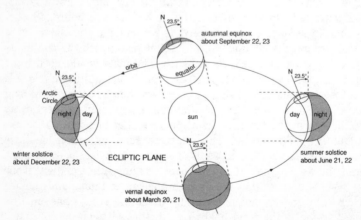

Figure 1.1 The orientation of the Earth's axis determines seasons, because as the Earth orbits the sun the duration of daylight varies.

Twice during the course of a year, around 21 or 22 June and 21 or 22 December, the axis tilts directly towards (or away from) the sun. These are the **solstices**, when it is midsummer's day and daylight is longest in one hemisphere, and midwinter with the shortest day in the other. The variation of one day is due to the difference between the calendar year and the slightly longer astronomical year, which is corrected for every **leap year** by adding a 29th day to February. This correction is a little too large; at the end of each century we normally miss a leap year to compensate. However that is also a slight overcorrection, so that every 400 years the century year *is* a leap year, including the year 2000.

Three months from each solstice the Earth's axis lies such that both poles are equidistant from the sun. These are the **equinoxes** when everywhere on Earth has approximately 12 hours of daylight and 12 hours of darkness. It is not *exactly* 12 hours for two reasons: first, the sun appears as a disc in the sky and not a point, producing a period of twilight when only part of it is below the horizon; second, the sun's rays are scattered by the atmosphere, especially when it is hazy, maintaining the glow of daylight beyond sunset. The **vernal equinox** occurs around 21 March and the **autumn equinox** around 21 September.

The tilt of the Earth's axis is the prime cause of seasonal changes of temperature and weather. These differences are, in general, greatest at the poles and least near the equator, although factors other than latitude come into play. Oceans are good at storing heat, so small islands have less extreme climates than places at the same latitudes in the interior of large continents. Nevertheless, the importance of day length can be appreciated by considering alternatives to our present situation. Just suppose the axis of the Earth was perpendicular to the ecliptic with no tilt, then everywhere on Earth would have 12 hours of daylight and 12 hours darkness throughout the year; every day would be an equinox and there would be no seasons. In contrast, if the axis lay within the ecliptic, mid-winter would bring darkness throughout the 24 hours for the whole winter hemisphere, not just polar regions. Summers would be much hotter, melting polar ice caps every year. Winters would be extremely cold, with the oceans freezing as far south as the United Kingdom and California in the northern winter, and around most of Australia and New Zealand in the southern winter.

The 23.5-degree tilt of the Earth's axis means that the highest latitudes at which the sun is directly overhead are 23.5 degrees north and south of the equator. These are the **Tropic of Cancer** and the **Tropic of Capricorn** respectively. Similarly the **Arctic Circle** and **Antarctic Circle** lie 23.5 degrees from the poles, that is at 66.5 degrees north and south of the equator respectively (see Figure 1.2).

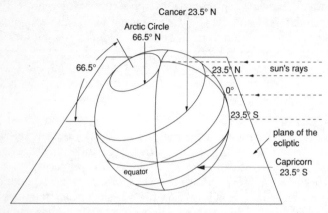

Figure 1.2 The Earth's axis tilts at an angle of 66.5° to the plane of the ecliptic, within which it orbits the sun.

These mark the boundaries from the poles of areas experiencing 24 hours of daylight in midsummer, or 'lands of the midnight sun', and 24 hours of darkness in midwinter. Even in midsummer the sun's rays are considerably weaker in polar regions than near the equator.

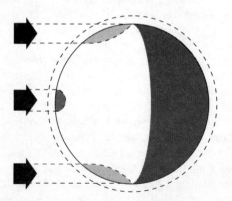

Figure 1.3 The angle of the sun in the sky dictates how much heat reaches the Earth.

This is not because of their longer journey through space which has no effect because there is nothing to weaken it, but first, because the sun's low angle near the poles causes its heat to be spread over a much wider area, and second, because more heat is lost over the longer path through the atmosphere (see Figure 1.3).

The geometry of the Earth, its orbit round the sun, its tilt to the ecliptic, and its daily rotation are all fixed, and all are crucial to our weather and climate. However, as we shall see, it is the Earth's atmosphere that provides the medium to enable our weather machine to function.

Latitude and longitude

Because the Earth is spherical, it is convenient to identify position on its surface by reference to circles, each defined with respect to its axis (see Figure 1.4).

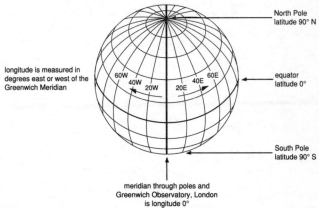

North Pole
latitude 90° N

longitude is measured in degrees east or west of the Greenwich Meridian

60W 40W 20W 20E 40E 60E

equator
latitude 0°

South Pole
latitude 90° S

meridian through poles and Greenwich Observatory, London is longitude 0°

Figure 1.4

Taking the Earth's axis vertically from pole to pole through the centre, the plane at right angles also passing through the centre of the Earth intersects its surface at the **equator**. A journey from equator to pole traverses 90 degrees of a vertical circle, and these are called degrees of latitude. Latitude is measured north or south from the equator; thus the poles are 90 degrees of latitude north or south, abbreviated to 90N or 90S; the tropics are 23.5N and 23.5S, and the Arctic and Antarctic circles are at 66.5N and 66.5S respectively.

Measurements east and west have conventionally used Greenwich, London as the zero point. If you imagine the Earth as segmented by vertical circles or **meridians** through the poles, rather like orange segments, then any point on the line from pole to pole through Greenwich is at 0 degrees of longitude. Positions east or west of the Greenwich meridian are measured in degrees up to 180 which is exactly halfway round. Thus a traveller going eastwards from Europe finds longitude increasing as he or she crosses Asia and proceeds over the Pacific Ocean to reach 180 degrees east, or 180E for short. This is halfway round the Earth and identical to 180W. Thereafter longitude is measured west of the Greenwich meridian, and decreases as the traveller journeys west to east across the Pacific Ocean, the United States and Atlantic Ocean back to Europe.

Degrees of latitude each represent an identical distance on the Earth's surface; degrees of longitude decrease considerably from equator to pole (see Figure 1.4).

2

THE EARTH'S ATMOSPHERE AND GENERAL CIRCULATION

The composition of the Earth's atmosphere

The **atmosphere** surrounds the Earth, both protecting it from harmful radiation and nourishing all living things. It is composed of **air**, which is simply a mixture of many gases. When the air is dry these gases occur in almost constant proportions across the Earth's surface; the main constituents by volume are given in Table 2.1. Oxygen is the prime requirement for us to live, with nitrogen and carbon dioxide particularly important for all plant life. The proportion of each constituent gas has changed slowly but considerably over geological timescales of millions of years, but until recently was believed to have reached a steady state. However, carbon dioxide has increased over the past century, and possible effects on future climate are discussed in Chapter 11.

Fortunately for us air is never completely dry, and by far the most important gas as far as weather is concerned is **water vapour**. This is a completely invisible gas, quite unlike clouds, mist or fog, which are composed of tiny liquid water droplets or ice crystals. The amount of water vapour in the air varies considerably in space and time, from less than one part in a thousand over arid deserts to around one part in twenty in hot and humid equatorial forests. It enters the air by evaporation from wet surfaces – notably from the oceans which occupy over two-thirds of the Earth's surface, but also from rivers, lakes and falling rain or snow. In addition it arises through transpiration from vegetation and, in comparatively small amounts, from the combustion

Table 2.1 The approximate composition of dry air [1]

Gas	Symbol	Percentage (by volume)	Cycle
Nitrogen	NO_2	78	Absorbed by growing vegetation. Released by microorganisms of decay.
Oxygen	O_2	21	Absorbed by humans, animals and fire. Released by plant growth.
Carbon dioxide	CO_2	0.05 (increasing see Chapter 11)	Absorbed by growing plants. Dissolves into oceans. Released by human and animal respiration, and combustion.
Argon	Ar	0.93	Non-reactive or inert gases.
Neon	Ne	Trace	
Helium	He	Trace	
Krypton	Kr	Trace	
Xenon	Xe	Trace	
Hydrogen	H_2	Trace	Very light; tends to be lost to outer space.
Ozone	O_3	Trace	Near the Earth's surface is a poisonous pollutant, but in the upper atmosphere shields us from harmful radiation – see Chapter 11.

Note: [1] Air also contains variable, usually small, amounts of other gaseous pollutants – mostly compounds of nitrogen and sulphur. These are influenced by the weather, and their importance is discussed in Chapter 11.

of fossil fuels and from the depths of the Earth in volcanic gases. It is removed from the air as rain, snow, hail, and drizzle, which are referred to collectively as **precipitation**. The unique characteristics of water in its three states of ice, liquid and vapour are discussed in Chapter 3.

The simple behaviour of gases

To understand why the atmosphere takes the form it does, it is essential to understand something of the way that gases behave. Every material is formed of minuscule identical building blocks called **molecules**, which themselves are composed of one or more **atoms** of basic elements. For example, a carbon dioxide molecule is composed of one carbon atom and two oxygen atoms, hence its chemical formula is CO_2.

Solid materials are composed of molecules locked together by strong attractive forces. The molecules maintain their positions relative to one another, and hence solid objects keep their shape, often despite substantial efforts to deform them. Liquids deform easily because molecular attraction is small enough to be overcome by their own weight. Consequently liquids 'flow' under the force of gravity taking the downhill path of least resistance, although they are prevented from doing so by a solid container. Whether flowing or not, the individual molecules are always in motion, continually changing position relative to one another.

Gas molecules by comparison have negligible molecular attraction to one another and are in continuous rapid, random motion. They proceed to fill the space available to them, readily mixing with other gases as they do so. This mixing process is known as **diffusion**. Unlike solids or liquids, gas molecules are widely spaced, consequently gases are easy to compress. A pneumatic tyre would burst before much more than its own volume of water could be pumped in. Gases are not weightless. The force of gravity holds the atmosphere down on the Earth's surface and prevents it drifting off into space, just as it does the oceans and every one of us! The weight of the atmosphere at any point is equivalent to about 10 m (33 ft) of water or 760 mm (30 inches) of mercury, and at any level the air is compressed by the weight of atmosphere above. Consequently, air near the surface is much denser or more concentrated than it is high above. For example at the top of Mount Everest at about 10 km (6 miles) a breath of air would take in less than one-third of the amount than at sea level. (The volume is assumed to be the same, but the weight and the **density** – weight divided by volume – would be much reduced.) This is why mountaineers need extra oxygen on high peaks, and why aircraft are pressurised to prevent the air within from expanding and making breathing more difficult.

The constituent gases of air vary little in proportion from the figures given in Table 2.1, with the exception of water vapour which varies greatly both in space and time. Water vapour is lighter than dry air so that, other things being equal, moist air is lighter or more buoyant than dry air.

Around 90 per cent of the atmosphere (by weight) lies in the lowest 15 km (9 miles) above the surface, and the Earth is about 12 640 km (7850 miles) in diameter. That makes our atmosphere, relatively speaking, a very thin skin indeed. To put it another way: we live at the bottom of an ocean of air which seems huge to us and is vital to our existence, but compared to the world itself it amounts to little more than a shallow if all-enveloping puddle!

The vertical structure
of the atmosphere

The atmosphere is made up of a number of layers each with different properties, temperature structure, and to some extent composition. None of these layers is absolutely constant in depth or character, but each maintains a large degree of consistency over the globe and through the year. The main features are shown in Figure 2.1.

The **troposphere** is the lowest region. It varies considerably and continuously in depth but on average is about 16 km near the equator and 9 km (5.5 miles) near the poles. It contains about 80 per cent of the total atmosphere by weight, and is by far the most important for our purposes because it is where virtually all weather occurs. Within the troposphere temperature falls with height in general, but the rate at which it falls, that is to say the **lapse rate**, varies considerably. Often an average value is quoted, but this can be misleading because the real lapse rate varies considerably in space and time. Sometimes the temperature is constant with height, in which case that layer is said to be **isothermal** ('iso' means equal or identical). Even more often, especially at night near to the ground, the temperature actually rises with height. In this case there is said to be a temperature **inversion**. Variations of temperature with height are, as we shall see, important factors in the weather machine. The upper boundary of the troposphere is called the **tropopause**.

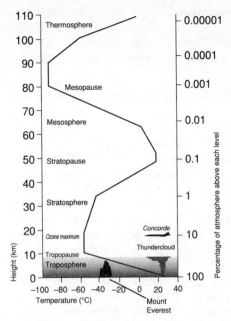

Figure 2.1 The temperature of the atmosphere. The temperature at any level varies day by day and season by season, but the average stays the same, and it is convenient to consider the atmosphere split into the distinct layers shown here. The lowest layer is the troposphere where all weather occurs. However strong winds in the stratosphere have an important influence on weather systems below.

The **stratosphere** lies immediately above the tropopause and extends to about 50 km (30 miles), where the boundary is termed the **stratopause**. Occasionally deep storm clouds penetrate into the lower stratosphere, but its main importance for us lies in the belts of strong winds that often occur in its lowest reaches and which are closely related to weather systems below. Most long-distance aircraft fly in the stratosphere above any adverse weather, taking advantage where possible of helpful tailwinds. The temperature structure in the stratosphere contrasts with that in the troposphere below. The temperature no longer falls with height, but is either constant or rises. This is because there is little vertical mixing of the rarefied air and it is heated directly by the sun. Since, by and large, the higher we go the more solar radiation is available for heating, the temperature rises accordingly. At the stratopause temperatures are roughly the same as at the Earth's surface.

The **mesosphere** extends above the stratosphere up to around 80 km (50 miles), and above that the **thermosphere** extends to beyond 300 km (180 miles). These will not concern us. Above the stratopause there is little air and no influences on our day-to-day weather.

—— Standard Atmospheres ——

For some scientific and practical purposes it is necessary or desirable to assume a consistent structure of temperature and pressure up through the atmosphere. For example, until recently most aircraft measured their altitude by **pressure altimeter**. This instrument relates the height of an aircraft above the Earth to the reduction in pressure between that at the level of flight and that at the surface. This difference depends on the temperature distribution through the atmosphere below the aircraft which cannot be measured and will almost certainly vary as flight proceeds. To overcome this and also ensure that such altimeters are consistent, the International Civil Aviation Organization (ICAO) has specified a **Standard Atmosphere**. This is a fixed and simplified average temperature profile through the atmosphere, and it is used to calibrate *all* instruments of this sort. It is shown in Table 2.2. It must be stressed that all such Standard Atmospheres, and there are

Table 2.2 The ICAO Standard Atmosphere

Height	Temperature
Mean Sea Level (MSL)	+15 °C
Temperature decrease of 6.5 °C per km through the troposphere	
11 km tropopause	−56.5 °C
Temperature constant at −56.5 °C in the lower stratosphere to a height of 20 km	
20 km middle stratosphere	−56.5 °C
Through the upper stratosphere an increase of 1.0 °C per km up to 32 km	
32 km stratopause	−44.5 °C

others, are hypothetical and only fleetingly exist in practice. Aircraft using pressure altimeters calibrated on the ICAO Standard Atmosphere do not register the aircraft's true altitude, but errors will be consistent between them. Therefore, when air traffic controllers assign different flight levels to aircraft they know that pilots will measure altitude using the same standard. Near the ground it would be dangerous to set altimeters in this way, because if the surface pressure happened to be particularly low an aircraft would be much closer to the ground than the altimeter indicated. In this case, the altimeter is always reset to the observed airfield pressure.

The general circulation
of the atmosphere

To many of us living in middle latitudes the day-by-day variation of wind and weather may seem almost random, and it could come as a surprise to learn that over most of the globe it is markedly different. This will become clearer as we discuss the main weather zones of the world.

Our weather engine is driven by heat from the sun. We saw earlier that heat reaching tropical regions is much greater than in high latitudes, because of the sun's height in the sky and the depth of the atmosphere the rays traverse (see Figure 1.3). The general circulation of the atmosphere acts to redress this imbalance, with warm winds spreading excess heat polewards, to be replaced by cooler air returning from the poles. If the Earth was stationary, these winds would be fairly direct: hot tropical surfaces would heat the lower atmosphere causing rising currents of warm air. These would flow outwards towards the poles as cooler, low-level winds blew equatorwards to take their place. The result would be a simple conveyor-belt type of circulation, known as a **Hadley Cell**, first explained by G Hadley in 1735. The real general circulation is much more complicated for several reasons, the main ones being seasonal changes in solar heating, and the Earth's rotation, size and geography.

The more complex actual general circulation can conveniently be subdivided into the four main zones, shown in Figure 2.2. They have a strong correlation with climates around the world, which are discussed in Chapter 12.

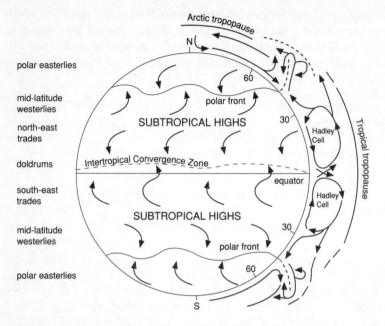

Figure 2.2 The major wind zones around the Earth result from excess solar heating in equatorial regions transported towards the poles, and winds which are strongly influenced by the Earth's rotation.

Equator to about 20 degrees of latitude – the tropical Hadley Cell

A Hadley Cell-type of simple wind system does exist, but only near the equator, extending north and south to about 20 degrees of latitude. The Earth is rotating from west to east, and consequently both the Earth's surface and the air above it are moving eastwards. Near the equator the air is moving more rapidly eastwards than air to the north (or south) because the radius of rotation is greater. Indeed, the speed in space of a point on the Earth's surface varies from zero at the poles to about 460 metres per second (m/s) (900 mph) at the equator. Any wind heading polewards from the equator maintains much of this eastwards velocity, while the surface beneath it is moving progressively more slowly as latitude increases. Thus, an initial southerly upper wind heading towards the North Pole becomes a south-westerly. In the same way, the returning low-level winds of the Hadley Cell, heading towards the equator, are deflected westwards to become north-easterly in the northern hemisphere, and south-easterly south of the equator.

Larger cells would be subject to greater deflections, the effect of which would be to set up large eddies which would stop the simple north–south circulation (see Chapter 4).

While Hadley Cells do not cover the globe, they strongly influence the major weather patterns over more than half the Earth's surface. There is a narrow zone near the equator where low-level north-easterlies from the northern hemisphere and south-easterlies from south of the equator meet, and the air is forced to rise. This is called the **Intertropical Convergence Zone** or **ITCZ** and is important because it often produces wide areas of thunderstorms and torrential rain, the lifeblood of equatorial forests. The ITCZ varies in width from a few kilometres to a few hundred kilometres. Sometimes it is so weak it is impossible to detect, even on satellite pictures; at other times it is clearly marked by large areas of deep cloud. It does not stay in one place but varies day by day and week by week; nevertheless it always tends to advance into the summer hemisphere where heat from the sun is strongest. Therefore, the Hadley Cells themselves are not constant, but edge furthest into the summer hemisphere. Over the sea the range of movement of the ITCZ is typically a few hundred kilometres, over the continents it is much larger reaching 2000 km (1250 miles) or sometimes more.

Low-level winds in and near the ITCZ are usually light, and over the oceans are known as the **doldrums**. They were dreaded by the crews of sailing ships because of the difficulty of making progress, compounded by heat and high humidity. The low-level, brisk, and nearly constant north-east and south-east winds blowing for hundreds of kilometres across the tropical oceans towards the ITCZ are known as the **trade winds**. This term also stems from the days of sailing ships which relied on them to make good progress. Over the continents in low latitudes the surface winds are much more influenced by geography, and neither doldrums nor trade winds exist in such clearcut form.

20–35 degrees of latitude – subtropical high pressure

In this zone the high-level outflow of air away from the tropics, comprising the upper limb of the Hadley Cell, descends. It is an area of large and remarkably persistent high pressure cells known as sub-tropical highs. These lie outside the tropics, sometimes extending beyond 35N and 35S especially in the summer, and usually have little or no cloud or rain, with light winds. The obvious consequence of this prolonged dry, clear and hot weather, is that continents in these latitude bands contain most of the world's deserts. On the other hand, islands and coastal regions which do enjoy occasional winter showery spells are not only highly productive agriculturally, but provide holiday paradises!

35–70 degrees of latitude – changeable westerlies

From about 35 degrees polewards to around 70 degrees of latitude lie broad zones within which westerly winds predominate, but these winds are often punctuated by interruptions as large eddies move around, producing variations in the weather. Sometimes dry or wet spells persist for weeks; at other times the weather seems to change every day. It is with unassailable logic that these are called the zones of changeable westerlies. Here, especially in the middle or on the eastern side of large continents, droughts are not unknown, while any location may expect severe weather from time to time. In general, however, climatic conditions are not extreme. Large weather systems, usually (but not always) travelling eastwards, bring rain and strong winds, interspersed with quieter areas of light winds and little cloud. The westerlies are strongest and most persistent over the open oceans and in the southern hemisphere became known to sailors as the **roaring forties**.

This temperate climate favours many crops and livestock, providing the wherewithal for human settlement. Furthermore, since many of our activities depend on the vagaries of weather, forecasting it has become particularly important. Weather systems and features of this zone of changeable westerlies comprise much of the content of this book. Because of the Earth's rotation we have seen that a high-level southerly wind heading northwards from the equator soon becomes south-westerly. If it retained all of its original energy or momentum, by the time it was halfway to the pole the wind speed would be more than 500 m/sec (1100 mph). In reality, long before such speeds can be attained, eddies form and the energy is diverted in developing large weather systems. It is these systems that serve to take warm air towards the poles through middle latitudes, and bring colder air from high latitudes to replace it. A major part of weather forecasting is determining where and when these systems will form and how they will develop and move.

70–90 degrees of latitude – polar high pressure

Polar regions are distinct from the zone of unsettled westerlies, because they see more easterly winds and pressure is often high. Winter cooling in polar regions generates a Hadley Cell-type circulation which can be quite persistent. As air cools through the long dark Arctic winter it becomes heavier or more dense than that further south. The result is that it sinks, flows southwards and is deflected westwards because of the Earth's rotation. Winds at higher levels are drawn towards the pole to replace it. This weak Hadley Cell pattern does lead to settled periods of weather in polar regions, but it is most unreliable. Often weather systems bring snow and strong winds, and at these times polar regions simply become extensions of the unsettled westerlies – but the climate is certainly not temperate!

In winter, high in the stratosphere, strong westerly winds develop due to intense cooling. However, this is far above the weather and not of direct importance.

It would be reasonable to expect a gradual decrease in temperature as we proceed polewards from the warm sub-tropical high, but this is not so. The deflection of winds by the Earth's rotation and other influences tend to concentrate the main temperature changes across relatively narrow zones, varying in width from tens to a few hundreds of kilometres. These are called **frontal zones** and can often be detected on weather charts by bands of cloud, sometimes extending thousands of kilometres round the hemisphere, varying in width, latitude and intensity. They frequently mark areas of rain or snow and strong winds, and are discussed further in Chapter 5.

3
THE WONDER
OF WATER

In our weather machine water is the most important component. It is essential for all forms of life, and the lack of it in animals or crops soon leads to dehydration, stress and eventually death. The atmosphere extracts water from the seas, oceans, forests and wetlands of the world, rain bearing clouds form and are often taken many thousands of miles to the interior of continents. There precipitation nourishes crops, fills reservoirs and lakes, and feeds the streams and rivers that eventually flow back into the sea. This complete process is called the **water cycle**, because it is continuously repeating. Having said that, there are great differences in the timescale: snow falling over Arctic regions may be consolidated into the ice field and remain locked there for thousands of years; at the other extreme a coastal shower may return water to the sea within an hour or so of the cloud forming. In this chapter we are concerned not only with the formation of cloud and rain, but also look more closely at the part played by water in each of its three **states** or **phases**: liquid water, solid ice and water vapour gas.

— Water as ice, liquid and vapour —

The water molecule comprises two hydrogen atoms and one oxygen atom and is consequently sometimes referred to simply by its chemical formula, H_2O. The attraction between adjacent H_2O molecules varies depending on their state. As **solid ice**, the molecules are almost stationary, locked in a lattice-like structure. Nevertheless, there is a continuous even if quite undetectable vibration which corresponds to their internal energy or heat; we measure this energy level as temperature. With **liquid water** the molecules have more heat energy, sufficient to enable them to escape the attraction of their neighbours and move about or 'flow'. In gaseous form as **water vapour** each molecule is much more remote from its neighbour and in constant rapid random motion. The energy level of an individual vapour molecule is considerably higher than that of one in liquid water or ice, but because they are much more spaced out the temperature of vapour can vary through a considerable range, and is not always higher than that of liquid water or ice.

Usually a frozen substance occupies less space than when it is in liquid form, but water is an important exception. Ice is lighter, or less dense, than the same volume of water, which is why icebergs float. In fact, water is at its most dense, or heaviest, at a temperature of near 4 °C; when cooled below that it starts to expand. This fortunate circumstance means that ice forms at the surface of rivers, lakes and ponds, allowing life to go on in the slightly less cold, unfrozen water below.

Latent heats

To change ice into water it is necessary to provide heat to raise molecular energy levels and break the bonds between. However, it is found that when heat is applied to ice and it melts there is no change in temperature. The heat energy that appears to be 'lost' is used up in breaking down the rigid lattice of ice molecules. Conversely, when water at freezing point turns to ice, heat energy is released even though the resulting ice is at exactly the same temperature. The heat involved is called the **latent heat of fusion** (latent meaning 'hidden'). In a similar way, heat is required to turn water to vapour with no change of temperature, and is released by the reverse process of condensation; this is called the **latent heat of vaporisation**. Much less common is the direct deposition of ice on to very cold objects from water vapour in the air, and the transformation of ice directly into vapour. Both processes are known as **sublimation** and involve the release or absorption of both the above latent heats.

Latent heats are not trivial. To melt ice into water at 0 °C takes the same heat as would be needed to heat that water from 0 °C to 80 °C, which is hot enough to scald. Water and water vapour are in equilibrium at its **boiling point** which is normally 100 °C. (The exact temperatures at which water freezes or boils vary with pressure. On a mountain top water boils below 100 °C and it is claimed that tea does not taste as good as at sea level!) To convert water to vapour at boiling point needs the latent heat of vaporisation which is enough to raise over five times the same amount of water from freezing to boiling point!

Changes from vapour to water to ice (and the reverse) play a vital part in the formation (and dispersion) of clouds and precipitation, because they involve the absorption and release of vast quantities of heat. Indeed, the release of latent heat provides much of the energy for weather systems themselves, especially hurricanes and thunderstorms.

Evaporation

We have seen that the heat energy of substances, and hence their temperature corresponds to the vigour of the motion of constituent molecules. When water reaches its boiling point of 100 °C, further heat will turn it to vapour at the same temperature. (Steam, which comprises condensed liquid water droplets must not be confused with water vapour which is an invisible gas.) However, water will slowly turn to vapour at much lower temperatures. Washing hung out in the wind is not boiled dry, and neither is a damp beach towel spread out to dry after a swim; the water **evaporates**.

At every water surface open to the air, some of the more energetic molecules escape. Each of these takes with it the required latent heat energy, and the consequential temperature loss of the water has to be made up by heat transferring in from outside. Usually this is from the air nearby or from sunshine. If we wear damp clothes, the heat needed for evaporation may be taken from our bodies and we soon feel cold, which is why waterproof outerwear is so important in cold, wet weather. A beach towel is made of cotton fibres with a large surface area. These not only provide an absorbent drying surface after our swim, but also allow evaporating water molecules to escape readily, helped by a breeze or direct heat from sunshine. Eventually our towel will feel completely dry, even though of course, its temperature never approaches 100 °C!

Humidity

Meteorologists need to know how much water vapour is in the atmosphere around us, or in other words the **humidity** of the air. That determines whether cloud or fog might form or if already present, whether they are likely to evaporate. It is a vital factor in deciding whether showers or storms are possible, and in predicting rain from large and small weather systems. The most fundamental measure of humidity is the **humidity mixing ratio**, which is simply the ratio by weight of water vapour to the dry air containing it, usually expressed in grams per kilogram (g/kg). Typical values in temperate latitudes are 10 g/kg, but it varies from near zero over deserts to 40 g/kg or more in equatorial rainforests and over tropical oceans. Clearly, as water evaporates, the humidity mixing ratio increases. Indeed the evaporation process itself provides the basis for the most common method of measuring humidity, described in Chapter 6.

There is one circumstance in which our damp beach towel will never dry, that is when the air already holds as much water vapour as possible. The air is then said to be **saturated**. In this case, any energetic water molecule escaping from a water surface is immediately replaced by another displaced from the air around it. In other words, evaporation is exactly balanced by the reverse process called **condensation**. Warm air can hold much more water vapour than cold air, but assuming no heat comes from outside the process of evaporation will cool the air into which it takes place. This is why a wind hastens drying. Escaping molecules and the air they cool are whisked away, to be continually replaced by drier air.

Most of us have experienced hot **muggy** days, so humid they make us uncomfortable. This is because evaporation of perspiration is an important mechanism to enable the human body to keep cool. When surrounding air is saturated this is impossible and we soon become hot and sticky.

Two further useful measures of humidity are commonly used by meteorologists, dewpoint and relative humidity. Often the absolute water content of air is of less interest than how much the air can cool before it becomes saturated, at which point cloud, or fog or dew may start to form. That critical temperature is called the **dewpoint**, and is included in all standard meteorological observations. If you breathe on to cool glass, moisture condenses on it as 'dew', because it cools the moist breath below its dewpoint. The dewpoint can never exceed the air temperature, but the two are identical when air is saturated. Whatever happens to the temperature of the air, provided its water content is unaltered then its dewpoint and humidity mixing ratio remain the same.

The warmer air becomes, the more water vapour it can hold. Often the unused capacity is of as much interest as the absolute amount of moisture held. This is given by the **relative humidity** which is the ratio of the amount of water air contains to the maximum it could possibly contain, expressed as a percentage. Absolutely dry air (which never occurs in nature) would have zero relative humidity, while saturated air (which is fairly common) has 100 per cent. Increasing air temperature will decrease relative humidity because the vapour carrying capacity is increased; cooling air increases its relative humidity until, at its dewpoint, 100 per cent or saturation is reached. Relative humidity is always close to 100 per cent in fog and cloud, while at the other extreme in the heat of the day over inland deserts it may fall to as low as 2 or 3 per cent. Strangely it is possible in special circumstances for it to exceed its normal 100 per cent maximum, a paradox that will be explained as we proceed to look at how clouds form.

— The formation of clouds and rain —

Usually temperature decreases with height through the troposphere. On the face of it, this is surprising. When air is heated it expands and becomes less dense, that is to say more buoyant, than its surroundings. Consequently it rises, after all this is how hot air balloons get off the ground. Why then doesn't rising warm air bathe mountain tops in balmy breezes, while far below we freeze in icy winds? The answer lies in the compressibility of air and its energy balance.

Most of the incoming heat from the sun arrives at the surface of the Earth raising its temperature, which in turn warms the air in contact with it. Consider a large bubble of air near the ground warmed by the sun. It expands and starts to rise. As it rises there is less atmosphere weighing down on it from above, that is to say the pressure on it decreases, and therefore it expands. This expansion takes energy, and the only immediate source of energy is within the bubble itself in the form of heat. Consequently, as the air rises some of this internal heat energy is used up, and the temperature of our bubble falls.

This type of process, where no heat is transferred in or out, is called an **adiabatic** change. In cloudless air the temperature fall with height is 9.8 °C/km (almost exactly 1 °C/100 m, or 5.4 °F/1000 ft) which is called the **dry adiabatic lapse rate** of temperature, or **DALR** for short. When our cooling bubble reaches the level where its temperature is the same as its surroundings then it stops rising. In a similar manner, descending air currents become warmer at the DALR as they are compressed by increasing pressure, which dissolves cloud. The sun's heat, therefore, is taken upwards through the atmosphere, but at the same time the lower atmosphere usually settles down to a temperature decrease or lapse with height.

Now consider another bubble starting from near the surface, which is warmed and starts to rise slowly through the troposphere, but this time comprising moist air. The amount of water vapour in the ascending bubble will not change; however as the air expands and cools at the DALR it can hold less and less, and its relative humidity increases, until it reaches a level at which it is saturated with water vapour. This is called the **condensation level** because continued ascent above this and further cooling leads to condensation of water vapour into tiny droplets and cloud starts to form. In fact, the droplets form around minute impurities in the air known as **condensation nuclei**, minute particles of dust or sea salt, much too small to be visible to the naked eye and normally present in large numbers. Condensation releases latent heat, warming the bubble, so that from this point on as the bubble rises the temperature loss is slower. This means that the relative buoyancy of the cloudy bubble is greater than an unsaturated bubble, encouraging further ascent and deeper cloud. Assuming as before that heat neither enters nor leaves from outside, temperature in cloud falls with height at the **saturated adiabatic lapse rate** or **SALR**. At low levels this is about half of the DALR, but as temperature and water content falls the two rates converge, so that near the tropopause there is little difference.

It is important to appreciate that cloud is not formed of raindrops. Cloud droplets are much smaller, initially only about 0.001 mm (0.00004 inches) across, rather less than the cross section of a human hair. Considerable growth is needed to produce even a small drizzle-sized drop – about 0.1 mm (0.004 inches) in diameter, let alone rain-drops which are 1 mm (0.04 inches) or more across.

Cloud droplets are not absolutely uniform, and because of this they drift within the cloud at different speeds. This leads to collisions between droplets, and **coalescence**, increasing the droplet size. **Drizzle** forms in this way, often starting to fall within a few hours of a cloud forming. However, raindrops are much bigger, and to produce a raindrop simply by coalescence would take several days. Since rain showers often fall from cloud less than an hour old, other processes are clearly at work.

Consider our cloudy bubble ascending further to where the temperature is below freezing. Cloud droplets do not freeze the instant the air temperature falls below 0 °C, because the latent heat released cannot easily be absorbed by the surrounding air. Until the cloud is colder

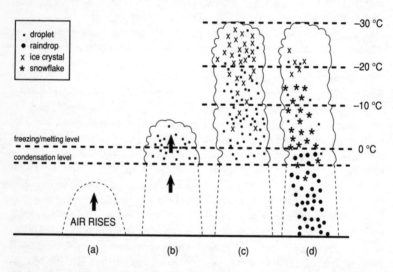

Figure 3.1 The development of cloud, and the formation of rain.

(a) As air rises, it expands, and cools as it does so.

(d) Above the condensation level droplets form to make cloud.

(c) When it rises further, to where temperatures are well below freezing, ice crystals form.

(d) These grow at the expense of droplets and combine with other crystals to for snowflakes, which fall through the cloud and usually turn to rain before reaching the ground.

than about –10 °C the cloud remains composed almost entirely of water droplets, and from –10 °C down to about –30 °C large quantities of droplets persist, although more and more ice crystals form. Droplets at subzero temperatures are said to be **supercooled**. Below –30 °C clouds comprise mostly ice crystals, and below about –40 °C no supercooled droplets remain. It is the coexistence of water droplets and ice crystals in cloud that leads to the development of individual large raindrops.

As rising cloud continues to cool below 0 °C the first ice crystals form on minuscule crystalline impurities in the air; usually of silica from dusty soil or salt swept up from the sea. It appears that the angular crystalline structure of these tiny particles mimics the lattice structure of ice molecules, which encourages freezing. Although these crystalline nuclei are minute they tend to be the largest (sometimes referred to as **giant condensation nuclei**) and most able to conduct away the latent heat as the surrounding droplet freezes. There now begins a critical stage in the lifetime of the cloud, now comprising a mixture of tiny droplets and ice crystals, when crystals grow rapidly at the expense of the droplets.

The air in the rising cloud is saturated as far as the water droplets are concerned. That is to say as fast as water molecules escape from a droplet surface (evaporate) a similar number are deposited on to it (condense). In an ice crystal, however, the molecules have much less energy and are held more rigidly. A few escape from each crystal (sublimate) into the surrounding air but are far outweighed by incoming condensing molecules from the surrounding water vapour, which immediately freeze. Each crystal gains many water molecules from the surrounding air and consequently grows rapidly in size. The air is said to be **supersaturated** with respect to ice. As ice crystals take H_2O molecules from the air, the air rapidly becomes drier, and therefore unsaturated with respect to the water droplets, which now start to evaporate. Thus we have an apparent contradiction – the air is supersaturated with respect to ice, but unsaturated with respect to water. The net effect is a rapid transfer of water to ice crystals which grow quickly at the expense of the surrounding droplets. This is the **Bergeron** (or **Bergeron–Findeisen**) process, named after the Norwegian who first explained it. It produces large ice crystals and eventually raindrops hundreds of times faster than the simple coalescence of droplets.

As ice crystals grow bigger and heavier they start to fall through the cloud, colliding with other crystals and further droplets which, at low temperatures, freeze on impact. Later, as crystals fall through warmer cloud, droplets do not immediately freeze on impact but act as a sort of cement, capturing other crystals and forming larger and larger snowflakes. This process is known as **aggregation**. In very cold weather **snowflakes** are small when they reach the ground, rather like sparkling dust, because the crystals tend not to stick to one another. When, more commonly, the falling snowflakes reach air above freezing they melt, and subsequently arrive at the Earth's surface as rain. In middle latitudes virtually all rain is melted snow, although sometimes heavy rain from storm clouds is melted hail.

The Bergeron rainfall process is by far the most common way in which rain forms, whether from huge weather systems or individual shower clouds. However, other processes occur in deep and turbulent cloud, to produce some of the heaviest rain and hailstorms.

Stability and instability: rain and showers

All precipitation forms from the ascent of moist air, but meteorologists distinguish between dynamic rain from large weather systems and showers which fall from individual cellular cloud. **Dynamic rain** (strictly dynamic precipitation) usually occurs over an area of hundreds or thousands of square kilometres and may persist for days, often moving several hundreds of kilometres in that time with its parent weather system. Cloud from which it falls has usually formed during slow ascent of air over a period of many hours or days. A **shower** on the other hand is usually of short duration, rarely more than half an hour, and falls from smaller and more short-lived cumulus or cumulonimbus cloud, although the air currents within them are often more violent. Showers present a different and often more difficult forecasting problem because of their size and transient nature.

Instability and showers

The bubble discussed in the previous section was assumed to be triggered by surface heating from the sun, and that is the way many showers form.

This happens only if the temperature lapse rate through the atmosphere will allow a bubble liberated in that way to rise high enough to produce cold cloud, ice crystals and rain, all before the cloud dissolves by mixing into its drier surroundings. The process is known as **convection**, and produces convective cloud (see Chapter 6). It will only occur when the atmosphere is warm and moist near the surface compared with deep cold air above. In this case, the atmosphere is said to be **unstable**.

Rarely, in extremely unstable situations, convective clouds reach the tropopause. Usually they are shallower because the unstable layer is rarely that deep, and besides rising bubbles tend to mix with their surroundings and lose impetus.

Air in convective clouds rises at speeds of typically 5–10 knots (kn), but this varies considerably both between clouds and within individual clouds. The deepest convective clouds are extremely turbulent with strong up and down currents, posing a threat to small aircraft. They are called cumulonimbus and are responsible for all thunderstorms and large hail. Clouds are discussed in more detail in Chapter 6.

In the most vigorous convective storms damaging hailstones may be produced. Rapid upcurrents feed moisture quickly to great heights and once ice crystals start to form they grow rapidly, not only through the Bergeron process, but also through copious collisions with droplets and other crystals. The largest soon become heavy enough to fall through the cloud and scavenge further drops and crystals by collisions, a process known as **accretion**. When further violent updraughts are encountered these large amalgamations are taken back up again into the top of the cloud, freezing as they do so, forming small **hailstones**. These are even more efficient scavengers of droplets and crystals which readily adhere to their surface, and soon the growing hailstones begin to fall again. Frequently, hailstones repeat this journey back up into the cloud several times before finally reaching the ground. Cross sections of large hailstones almost always show them to be layered, rather like an onion, clear ice being interspersed with opaque layers. The clear ice corresponds with impacts with large drops in the cloud, which spread round the hailstone before freezing, while the opaque layers, appearing like frosted glass, are where ice crystals have adhered to the wet surface without melting, trapping air pockets amongst them. Large hailstones are invariably associated with deep and vigorous convection, usually accompanied by thunder and light-ning which are discussed in Chapter 10.

Mass ascent and rain

Other mechanisms come into play to force air to rise much more slowly but over a much wider area, often many thousands of square kilometres. This external forcing can take the form of low-level winds gradually blowing inwards or converging, so that the air where they meet is slowly pushed up. Alternatively (or in addition), the normally much stronger winds high in the atmosphere may blow outwards or diverge, so that air beneath is sucked up. These processes of **convergence** and **divergence** of windflow are the basic processes associated with the huge areas of cloud and rain of large weather systems, discussed in Chapter 5. Because upward motion occurs over a wide area it is called **mass ascent**. An important point here is that the rate of ascent is slow – typically less than 1 kn, contrasting with horizontal wind speeds which even near the surface often reach more than 30 kn. Nevertheless, mass ascent often continues for many days. In these circumstances the air is said to be **stable** when without the external forcing it would maintain its current vertical temperature structure, and no further cloud would form. The processes leading to mass ascent and the development of large weather systems are greatly influenced by the Earth's rotation, which we will consider further in Chapter 4.

4

THE EARTH'S RADIATION BALANCE AND ROTATION

Weather is closely bound up with energy transfer and interchange between the Earth and the atmosphere. In Chapter 2 we saw that excess heat energy from the sun in tropical regions is spread by the winds of the general circulation of the atmosphere towards the poles, and there is a net flow of cooler air in the opposite direction. In this chapter we are going to look a little more closely at how heat from the sun reaches the atmosphere, and how that heat energy produces large movements of air. We will also consider the way in which the winds are greatly modified by the Earth's rotation.

The heat balance of the Earth and atmosphere

Heat from the sun reaches us in the form of electromagnetic waves which travel through space. Our eyes respond to light which forms a large part of the sun's radiation, although it is only a small part of the complete spectrum of electromagnetic radiation shown in Figure 4.1. This is likely to be a consequence of our evolution rather than a coincidence. The remainder of the spectrum is not visible to the naked eye: longer wavelengths include radio, television and infrared; shorter wavelengths contain ultraviolet rays that produce sunburn, along with X-rays and gamma rays.

Everything radiates electromagnetic energy, with an intensity and wavelength that depend critically on its surface temperature.

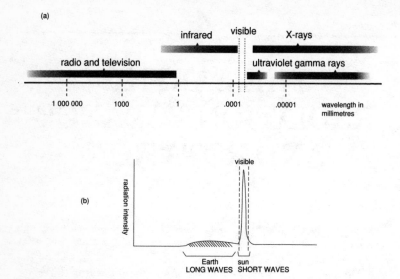

Figure 4.1 Electromagnetic radiation.

(a) Visible light is only a small part of the electromagnetic spectrum, but of vital important to us. The atmosphere largely protects us from damaging ultraviolet rays.

(b) Incoming radiation from the sun is highly concentrated about visible wavelengths, hence the sharp spike. Outgoing radiation from the much colder Earth and atmosphere is spread more thinly across longer wavelengths.

How heat is transferred (see Figure 4.2)

Conduction is the transmission of heat by the surface-to-surface contact of objects with different temperatures, such as by putting warm hands on to cold glass or metal.

Convection is the spreading of heat through a fluid when heating (or cooling) is applied at only one part, and the increased (or decreased) local buoyancy causes mixing. Examples are heating a pan of water on a hot plate or cooling a jug of water in a refrigerator.

Radiation is transmission of heat by electromagnetic waves, such as from the sun or a red-hot electric element.

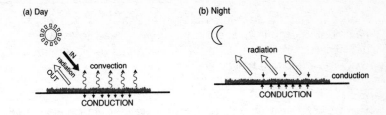

Figure 4.2

(a) Daytime radiation from the sun heats the ground. This warms the air next to it which spreads the heat upwards by convection. Some also travels down into the Earth by conduction. Although the Earth also radiates heat away, this amount is dwarfed by that coming from the sun.

(b) At night the Earth's surface continues to radiate heat away into the atmosphere and space, cooling in the absence of heat from the sun. Some of the heat lost is replaced by conduction of heat up from the ground, and down from the lowest layer of air.

Stefan-Boltsman's law says that energy radiated by a perfect radiator is proportional to the fourth power of its temperature. That is to say a doubling of temperature (measured in degrees Absolute) leads to a sixteen-fold increase in radiated energy.

The Earth and its atmosphere radiate as much energy back into space as is received from the sun; if that were not the case, then it would become hotter and hotter. There is an important difference between incoming and outgoing radiation: the sun, which has an average surface temperature around 6000 °C, transmits on considerably shorter wavelengths than the much cooler Earth. Furthermore, the sun's short waves are much better at penetrating through our atmosphere, though by no means all arrive at the Earth's surface.

Of all the radiation reaching the outer atmosphere from the sun, on average only about 50 per cent reaches the Earth, the rest being reflected directly back into space, absorbed by gases in and above the stratosphere, or intercepted by clouds and particles in the troposphere. The sun's rays that are not reflected are absorbed at the Earth's surface, raising its temperature and evaporating water. This

> **Wien's law** states that the variation of wavelength at which maximum energy is emitted is inversely related to temperature. That is to say a doubling of temperature (in degrees Absolute) leads to a halved wavelength.

portion varies greatly from place to place depending on the character of the surface; the proportion of incoming radiation reflected is known as the **albedo** of that terrain. Over forest or wet ground it can be as low as 5 per cent; it is around 20 to 30 per cent over dry earth and sand; and up to 80 per cent over fresh snow which is blindingly reflective. The albedo of water depends how high the sun is in the sky; it ranges from only about 5 per cent when almost overhead, to near 70 per cent when the sun is at a low angle.

The Earth's surface re-radiates energy on longer wavelengths, but nowhere near sufficiently to balance the incoming solar heating. Most of the excess is used either to evaporate water (supplying the latent heat discussed in Chapter 3), or to heat the air in contact with the ground. This warmer and often moister air mixes with cooler air above by convection currents and also wind-driven turbulence. In that way the sun's heating is spread gradually upwards throughout the troposphere. It is the total outgoing longwave radiation from clouds and atmospheric gases, added to that from the earth's surface that balances incoming shortwave solar radiation. Water vapour is a particularly efficient radiator. The important point as far as weather is concerned is that our atmosphere (strictly speaking the troposphere) is almost entirely heated from below by the Earth, not from above by the sun; it is the weather machine that has a crucial role in redistributing it. The overall radiation balance is shown diagramatically in Figure 4.3(a).

So far we have considered the average balance between incoming and outgoing heat energy, without distinguishing between the seasons or day and night. Clearly there is a great difference. On a sunny day, especially in summer, incoming heat from the sun far outweighs outgoing longwave radiation; there is a net heat gain and the temperature rises. On the other hand, at night solar heating is completely cut off, while the Earth, clouds and the atmosphere continue to radiate and lose heat, consequently the temperature falls. Typical temperature cycles through the 24 hours for summer and winter are shown in

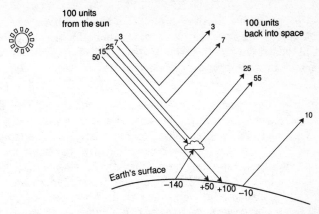

Figure 4.3 (a) Approximate average radiation balance of the Earth and atmosphere.

Of every 100 units from the sun: 3 absorbed by ozone in the stratosphere and re-radiated back into space, 7 scattered by the atmosphere and lost back to space, 25 reflected by clouds back into space, 15 absorbed by clouds and atmosphere, 50 penetrate the atmosphere and reach the Earth.

At the earth's suface: 50 units from the sun, 100 units from the atmosphere and clouds, 140 units lost into the atmopshere, 10 units lost directly into space.

In the atmosphere (including clouds): 15 units from the sun, 140 units from the Earth, 100 units lost to the Earth, 55 units lost directly into space.

Returned to space 100 units: 25 from cloud by reflection, 3 radiated by high-level ozone, 7 scattered by atmosphere straight back into space, 55 radiated by clouds and atmospheric gases, 10 radiated directly by the Earth.

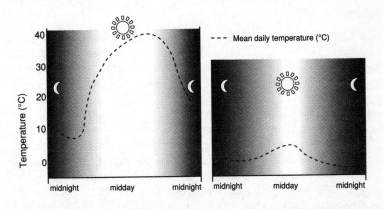

Figure 4.3 (b) Surface temperature through the day in summer and winter.

Figure 4.3(b). These will be modified considerably if there is cloud to obstruct the sun during the day, or provide a blanket above the Earth overnight. If it is windy the air is stirred up and any heating or cooling will be mixed through a much deeper layer of air than in calm conditions, and any temperature change at the surface will be much smaller.

In the summer under clear skies, outgoing radiation during the short night cannot possibly balance the heat arriving throughout each long, hot day. This might be expected to lead to progressively higher temperatures, and this does often happen for a while. Two balancing factors come into play: first, the temperature increase boosts outgoing longwave radiation both by day and night; second, the higher surface temperature causes the depth through which the air mixes by convection to increase. Over hot deserts this convective mixing layer sometimes extends from the surface to over 5000 m, where high level winds remove the rising heat. This layer has a near dry adiabatic lapse rate (see Chapter 3).

Long, calm winter nights in high latitudes lead to very low surface temperatures, especially where the air is too dry for a protective blanket of fog to form. Daytime sunshine is too weak to compensate. In this case, balance is also eventually reached: partly by downward longwave radiation from the atmosphere above, which is soon markedly warmer than the thin layer in contact with the ground; partly by reduced outgoing radiation from the colder surface; and partly by heat conducted upwards from below the surface of the ground.

Most land surfaces, especially when dry, are poor conductors of heat. Consequently, their temperature rises rapidly in the sun, and falls quickly after dark on clear nights. By contrast, the temperature at the surface of the sea varies little between day and night. This is not because water is a better conductor; first, the sun's rays penetrate well below the surface and spread their heat over a considerable depth; second, cooling at the surface mixes downwards by convection; third, it takes more heat to raise the temperature of water than the same amount of rock or soil. This marked difference in the response of sea and land surface temperatures to external heating and cooling is extremely important. It affects climate, the development of weather systems, and, as we shall see, it is responsible for sea-breeze and monsoon wind circulations.

Radiators

The efficiency of an object as a radiator of heat (strictly electro-magnetic waves) depends greatly on the character of its sur-face. Dark surfaces are generally the most efficient in sun-shine, warming more quickly and cooling down faster than white surfaces. Because of this scientists call a perfect radiator a **black body radiator**; this absorbs all incoming radiation and emits the maximum possible.

Few scientific terms are more misleading. The sun is an almost perfect black body by this definition, and so is freshly fallen snow to long waves, even though it reflects about 80 per cent of visible solar radiation. Because of this, when skies clear fol-lowing snowfall the temperature plummets, and the lowest tem-peratures on Earth occur over snow-beds in Arctic regions.

—— **Effects of the Earth's rotation** ——

We have seen that heat from the sun reaches the atmosphere largely by way of the Earth's surface, and that it is distributed around the globe by the winds and weather systems of the general circulation. The reason why winds do not simply take the excess heat in tropical areas directly towards the poles in the simple Hadley Cell-type circu-lation introduced in Chapter 2, is primarily due to the Earth's rota-tion. This is an important component of our weather machine.

The effects of rotation may be regarded as being composed of two quite distinct components. One is concerned with the conservation of energy. The other is to do with our frame of reference, the fact that we measure winds and discuss forces which act in straight lines with reference to a spherical world which rotates along with all of us and our measuring systems.

The conservation of energy

We are rarely if ever directly aware that we live on a rotating sphere, the progression of day and night is quite taken for granted. Yet those living on the equator, furthest from the Earth's axis, are effectively travelling through space at about 460 m/s (1030 mph). By contrast near the poles, where the radius of rotation is small, there is little speed due to rotation. As air flows towards higher latitudes then its radius of rotation decreases. This energy due to rotation is called **angular momentum**, and in any motion energy must be conserved. This means that the speed of air moving polewards must increase, to compensate for the reduced radius of rotation. In reality other factors come into play; the motion is diluted by mixing with surrounding air and slowed by frictional forces. Winds blowing equatorwards, on the other hand, find the Earth's surface beneath them proceeding progressively faster eastwards. They may pick up some of this momentum, but even so usually become north-easterlies.

The general westerly winds of middle latitudes, which mostly steer weather systems eastwards, are the result of large masses of warm air moving polewards injecting westerly movement into the flow at the new latitude. The effect is greatest at middle and high latitudes where the radius of rotation changes most rapidly with northward or southward shift.

Frames of reference – the Coriolis effect

The Earth's rotation has another important effect on all the winds we measure. We can see this more easily if we consider a large disc or turntable rotating anticlockwise, just as if we were looking down on the Earth from above the North Pole (see Figure 4.4). If two players A and B are standing on this turntable, and A throws a ball towards B, the ball travels in a straight line. However, by the time it reaches B both players will have moved, A to A^1 and B to B^1. To both of them, unaware of their movement, it will appear that the ball has been deflected to the right and taken the curved path from A^1 to B. A similar apparent deflection will occur wherever they stand, and is greatest when they are near the centre of rotation. If the players stand on the perimeter (the equator) the effect is small; if they face each other across the axis (the pole) it is greatest.

Our artificial scenario is analogous to winds flowing over the Earth. The fictitious but extremely practical force is called the **Coriolis force**. It is of fundamental importance to wind flow around weather systems, except near the equator.

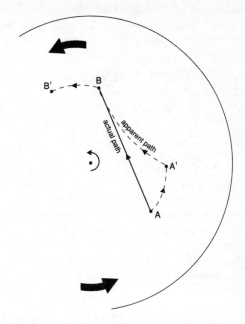

Figure 4.4 The Coriolis effect.

Geostrophic and gradient winds

Atmospheric pressure is simply the weight of atmosphere above any point. If the Earth was flat and did not rotate, wind would simply flow from high pressure to low pressure under the force of gravity. Imagine a pyramid of golf balls, the weight of those bearing down in the centre – or the high pressure – would soon force those near the ground outwards and the heap could not be maintained.

The force involved in the atmosphere is called the **pressure gradient force**, because it, and hence the wind speed, depends on the pressure difference or gradient between two points. Unfortunately wind is not as simple as that because, unlike the sudden and short-period collapse of a heap of golf balls, the longer timescale of winds means the Earth's rotation must be considered. The Coriolis force in the atmosphere acts to deflect the wind to the right. If just the pressure gradient and Coriolis forces acted, once the air started in motion it would curve to the right until they exactly balanced one another. At that point the air motion would be of constant velocity with low pressure to the left and high pressure to the right.

Buys Ballot's law states that if you stand with your back to the wind in the northern hemisphere, low pressure lies to your left and high pressure to your right. The reverse applies in the southern hemisphere.

This is usually a fair approximation of what actually happens, and explains why on most weather charts wind arrows lie more or less along lines of constant pressure or isobars. Scales have been derived to enable estimates of wind speed to be read from such charts, using the spacing of the isobars (measuring the pressure gradient) and the latitude (governing the Coriolis force). This estimate is called the **geostrophic wind**. In practice, additional smaller forces are involved, the most important of which are those due to friction and to the curvature of isobars.

Air flowing across the ground with its vegetation, buildings, hills, etc. is always slowed by friction (sometimes called 'drag'). Induced eddies often extend a kilometre or more above the surface, but the greatest effect is felt near the ground. Since the wind speed is reduced so is the Coriolis force; hence the balance of forces is altered in favour of the pressure gradient force, with the result that the wind blows slightly across the isobars towards low pressure and away from high pressure. Thus the geostrophic wind is a better approximation to the wind just above this so-called **turbulent boundary layer** where friction has no effect, rather than at the surface.

Wind varies considerably with height, as well as in space and time, usually increasing from the surface to the upper troposphere, but there is considerable variation. In the lowest few hundreds of metres friction ensures the wind rarely decreases with height except briefly near showers or thunderstorms. The loss of speed near the ground is partly compensated by overturning and mixing with the less affected layers above and, near the ground, wind flow is composed of a series of gusts and lulls due to this turbulent mixing. Gustiness is most apparent on a showery day when up and down currents are largest. At night the ground and the air next to it cools and the boundary between this cold blanket and the warmer air above is often quite abrupt. The 'surface' wind, or a large part of it, often blows above this cold layer, as if it was a slippery skin. Thus the night-time wind at ground level is usually much lighter than during the day. Aviators should be aware that they may suffer a sudden loss of lift as they descend through sudden wind changes.

The effects of curvature on wind flow are more subtle, but the result is simple. Around an area of low pressure or a depression the winds are somewhat less than geostrophic; while around high pressure, or an anticyclone, they are greater than geostrophic.

For anything to move in a curve a force is required directed towards the centre of the curve, and air moving over the Earth is no exception. All the force towards the centre of low pressure is due to the pressure gradient, so the inward force needed to make the wind curve has to be taken from this pressure gradient force, and the wind speed is consequently reduced. In an anticyclone the pressure gradient force is acting outwards (always towards low pressure), and in this case the extra force needed for the air to take a curved path must be taken from the Coriolis force. However, the Coriolis force must still counter exactly the unaltered pressure gradient force, and that can only happen with an increase in wind speed. Winds taking into account curvature of flow are called **gradient winds**, and the difference between gradient and geostrophic wind is sometimes called the **ageostrophic wind component**.

The above discussion does not apply to areas near the equator where, as we have seen, the Coriolis effect hardly exists. Indeed, there it is impossible to construct pressure charts with isobars, and meteorologists show the wind flow patterns directly as **streamlines**. An interesting consequence of the lack of geostrophic wind is that when small vortices such as dust devils or waterspouts occur at low latitudes, they may rotate in either direction even though central pressure is low.

Despite all these complexities, a simple geostrophic scale enables meteorologists to estimate winds with reasonable accuracy from weather charts. Computer programs can take all the variables into account much more precisely, and forecast winds for shipping and especially aircraft are now almost always taken directly from computer-produced forecasts.

The concentration of spin by convergence

The Coriolis force enables the effect that the Earth's rotation has on winds over the globe to be taken into account. It is of importance only when dealing with large-scale motion taking place over tens or hundreds of kilometres and lasting for several hours or days. Over the much smaller radius of a tornado, for example, which is usually less than 200 m (650 ft), and even more so the small vortex formed as water drains through a plug hole, the accelerations and forces of rotation make any contribution from the Coriolis effect completely insignificant. Tornadoes and water running out of a bath mostly rotate cyclonically (i.e. anticlockwise in the northern hemisphere), but not always, depending on pre-existing spin in the low level air, or the bath water. This spin is usually not apparent at all to an observer until it becomes concentrated: in the case of a tornado as the air converges to the small area where it is being sucked up into a thunder cloud; in the case of draining water as it converges towards the plug hole. This process of convergence increasing spin is of immense importance in the development and decay of large weather systems.

Most satellite pictures show one or more whirls of clouds, sometimes 1000 km (620 miles) or more across, marking winds spiralling into a depression. These huge systems are invariably formed by processes which concentrate spin through a large depth of the troposphere. The same principle is used by spinning ice skaters; they introduce rotation first with arms and often one leg outstretched, and then spin faster and faster by drawing their limbs close to their body, so concentrating spin over a smaller diameter.

Everything, including the air, has cyclonic spin because the Earth itself rotates cyclonically. Winds being drawn into an area by upcurrents always concentrate pre-existing cyclonic spin. This is fundamental in the development of hurricanes at low latitudes, and also depressions in middle and high latitudes. On the other hand, the divergence of air because of down currents can only proceed until the local anticlockwise or anticyclonic spin exactly counterbalances the cyclonic component due to the Earth's rotation. This is the absolute zero as far as spin on the Earth is concerned; no further reduction is possible, and this is why there are never vigorous anticyclones.

Jetstreams and Rossby waves

In the troposphere and low stratosphere above 8 km or so, belts of strong winds, 100 to 400 km (60 to 250 miles) wide and often many thousands of kilometres long, meander around both hemispheres through a series of loops or waves. These **jetstreams** were dreaded by early high-flying aviators because unexpected headwinds increased fuel consumption alarmingly. Now that jetstreams are accurately forecast, they can be avoided as headwinds and exploited as tailwinds, by routing aircraft accordingly.

At almost every point the wind blows along a curved path in space, even though to us it often appears to be in a straight line. In other words, the air has a degree of spin about the vertical. That spin has two parts to it: the first is due to the Earth's rotation at that particular latitude, which is a known fixed quantity; the second is relative to the Earth itself, which we can deduce from observations. As winds curve around the globe it is clear that the latitude, and therefore that part of the spin changes. However the total spin stays the same, just like the ice skater mentioned previously. That means the other part of the spin, that relative to the Earth, changes to compensate, and hence jetstreams curve and meander.

Meteorologists call spin in the atmosphere **vorticity** and the total spin **absolute vorticity**. Like energy and mass, vorticity is not created or destroyed in a free-flowing fluid, and this principle is known as the **conservation of absolute vorticity**.

Suppose at some middle latitude a stream of air high in the atmosphere is flowing north-eastwards away from the equator (see Figure 4.5(a)). Because the radius of rotation of the Earth decreases with increasing latitude, its spin due to the Earth's rotation increases. In that case, the component of spin relative to the Earth decreases to compensate. The net result is that the air turns clockwise, or anticyclonically, until it is heading south-eastwards; then the reverse occurs, and relative to the Earth it straightens up again. Once it overshoots its initial position, the Earth's radius of rotation is increasing and that part of the spin due to the Earth's rotation decreases (remember the skater?). Again, to compensate, the spin relative to the Earth must increase, and the air curves eastwards anticlockwise or cyclonically.

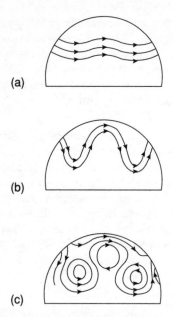

(a)

(b)

(c)

Figure 4.5 Rossby waves and the index cycle.

(a) The mobile or high index phase.
(b) The low index phase.
(c) Occasionally large Rossby waves break down or disrupt into separate circulations, with no net west to east movement. This is called a blocked phase, because the progress of weather systems is halted and the weather can remain little changed for weeks.

This process results in jetstreams meandering in a pattern of waves around each hemisphere, the waves themselves travelling from west to east. These are called **Rossby waves** after the scientist who first explained the process. They are of direct interest not only to the operators of high-flying aircraft, but also greatly to meteorologists, because the waves play an important part in the development of major weather systems.

Laboratory experiments and mathematical investigations have shown that the motion of a fluid in a slowly rotating container can be chaotic, but waves or oscillations that develop usually change only slowly. The Earth's atmosphere on our rotating world behaves similarly. This is little different from many things with which we are familiar – a pendulum, waves on the sea, the bough of a tree in the wind, or even an insect settling on a blade of grass. The oscillations are sometimes small, sometimes large. This also applies to the jetstreams flowing through Rossby waves high in the troposphere. They circle the globe in a gradually changing series of oscillations. When the waves are large and slow moving (see Figure 4.4 (b)), it is termed a **low index** phase and is often associated with vigorous, slow moving weather systems. On the other hand in a **high index** phase the north–south movement is small, and weather systems tend to move quickly and develop less (see Figure 4.5 (a)).

The energy from these powerful upper winds leads to the development of weather systems, some of which produce large areas of wet and windy weather that are so much a part of our weather machine. We shall learn more of these in Chapter 5.

5

WEATHER SYSTEMS

In our discussion of the weather machine we have seen that the general circulation transports heat polewards from the equator, with the resulting winds affected greatly by the Earth's rotation. In particular, powerful jetstreams develop high in the troposphere and meander around the globe. We have also seen how rain forms as moist air rises to where first cloud droplets and then ice crystals develop. Finally we saw how winds spiral into low pressure areas, rather than rushing straight in and filling them. Now we will fit these components together and look at the various types of weather systems; it is these that largely dictate where most rain falls, and where the sun shines.

Temperature changes, pressure and wind

Atmospheric pressure at the Earth's surface is simply the weight of the air above. If air is heated from below then it expands, becomes more buoyant, and warm currents rise and mix with the air above. Compared with adjacent unheated air the pressure at the surface will initially be unaltered, because the same amount of air remains in the column above. At a point higher up the situation is different. More of the air in the warmer column lies above that level than in the surrounding cooler zone. That is to say the pressure is higher, and air starts to flow out. This soon reduces pressure at the surface, and air starts to flow in. This is shown in Figure 5.1. If the heating is over a large area, the effect of the Earth's rotation will come into play, and

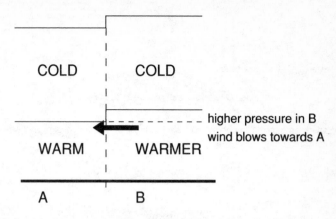

Figure 5.1 How temperature changes induce pressure differences and hence winds.

the Coriolis force (see Chapter 4) will cause the wind to blow anti-clockwise or cyclonically around low pressure.

This simple mechanism whereby temperature differences cause pressure changes and hence winds, forms the basis of almost all wind and weather systems, and is fundamental to our weather machine.

The sea breeze and land breeze

Perhaps the most common and certainly one of the simplest weather systems is the **sea breeze**. On a sunny day near the coast, heat from the sun raises the temperature of the land surface much more than that of the sea. The hot land surface warms overlying air which rises. Consequently, as explained above, the pressure falls over the land and, above the lowest kilometre or so, there is an outflow of air towards the sea. At low levels this outflow is replaced by a cooler wind from the sea which is the well-known sea breeze (see Figure 5.2). This is a common feature all over the world where land meets sea, especially where daytime temperatures are high and the sea cold in comparison. The low pressure over land caused by the difference of heating is called a **heat low**, and is marked over most hot countries through their summer. A sea breeze is usually most pronounced and common within 20 km (12 miles) or so of the coast, but during long, hot summer days can penetrate 100 km (60 miles) or more inland, bringing relief from sweltering heat. Its onset can be quite sudden; one

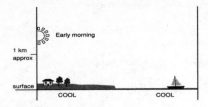

(a) Early morning: the temperature is similar at all levels above land and sea, and there is no wind.

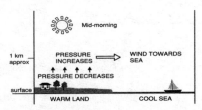

(b) By mid-morning the sun has raised the land temperature, so that higher up the pressure is increased and winds start to blow out towards the sea.

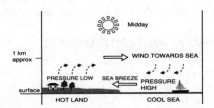

(c) Surface pressure over land is reduced, and the sea breeze blows in to compensate.

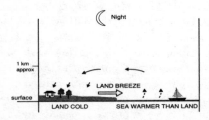

(d) At night the land cools more rapidly than the sea, and a weaker land breeze develops, blowing from land to sea.

Figure 5.2 Sea and land breezes.

minute nothing stirs, and the next trees are rustling and doors banging. The narrow zone where the wind changes is called a **sea breeze front**; it can be used by glider pilots to obtain lift on otherwise poor days for the sport, and sometimes birds are seen to find a ready supply of insects wafted high into the air by upcurrents. Occasionally a sea breeze front is marked by a line of usually shallow convective clouds (fair weather cumulus – see Chapter 10) where the cool, moist air from the sea pushes up against hot, stagnant air, causing it to rise above its condensation level.

Less well known is the opposite wind – the **land breeze**. This is a low-level flow of air from land to sea, generated during the night when the land has cooled to well below the sea temperature, which hardly changes. The land breeze is rarely as marked as a daytime sea breeze and is shallower, not only because the land to sea temperature contrast is smaller, but also there is no incoming solar energy to produce convective air currents and extend the circulation upwards. Unlike the sea breeze, the land breeze is most important in winter; when on a cold, still night it can provide enough mixing to prevent frost near the coast, and on a foggy night may spread fog across coastal waters.

In the latitudes of the British Isles the sea breeze rarely exceeds 10 kn and the land breeze is usually less than 5 kn; however they are rarely completely on their own. Although they occur in fine weather, there is usually a light gradient wind, and this combined with the sea breeze may total 15 kn to 20 kn. This can be uncomfortable, especially when low cloud or fog is brought inland, which often happens along the east coasts of England and Scotland. The resulting damp, drizzly weather from the cold North Sea is known as **haar** or **sea fret**.

In the tropics and subtropical high pressure zones where the sun is most powerful, sea breezes often reach 15 to 20 kn, especially where the land/sea temperature contrast is accentuated by cold coastal currents.

Detailed forecasts of the vagaries of local winds may be provided by meteorological services to benefit sailors in coastal waters, surf boarders and sun seekers. The forecaster takes into account the shape of the coastline, sea temperature, time of day and weather situation. In general, sea breezes are stronger in the afternoon and on to headlands and lighter into bays, and are little influenced by the Coriolis effect because of their local nature.

—————— **Frontal depressions** ——————

A sea breeze circulation is local, short lived and not associated with bad weather. Nevertheless the reduction of pressure and resulting wind circulation is replicated in a much more vigorous way in large weather systems. By far the most important of these in middle and high latitudes are frontal depressions.

A **depression** is an area of low atmospheric pressure, often simply called a **low**. Active, deepening depressions are associated with large areas of strong winds blowing anticlockwise around the centre, and wet weather. They are usually more or less circular and measure from about 500 km (310 miles) to 1000 km (620 miles) across. **Frontal depressions** form where there are large temperature differences through a large depth of atmosphere along a line or weather front, which may be thousands of kilometres in length; these are discussed in more detail later. The huge amounts of energy in a depression are obtained from three main sources: first, from the temperature difference across the front, in much the same way as the sea breeze; second, from the release of latent heat as massive volumes of cloud form in rising moist air; and third, from the strong winds, or jetstreams in the upper troposphere and lower stratosphere.

A depression forms where upper winds are taking away more high-level air than they are bringing in (see Figure 5.3 (a)); they can be regarded as causing a partial vacuum. As a consequence, the weight of air at the surface (or surface pressure) is reduced and winds at low levels flow inwards to redress the balance. We have seen how these are affected by the Earth's rotation, and deflected by the Coriolis force (see Chapter 4) into the familiar cyclonic circulation. The important consequence of this inflow or convergence at low levels and outflow or divergence aloft, is that air slowly ascends over the depression through a large depth of the troposphere. It is this mass ascent of air that causes the formation of wide areas of clouds and rain, and is why frontal depressions are invariably associated with bad weather. Cloud formation releases vast

---

Low pressure is not exclusively associated with bad weather. Heat lows often have hardly a cloud in the sky, and **lee lows** which form downstream of hills and mountains, often have clear skies.

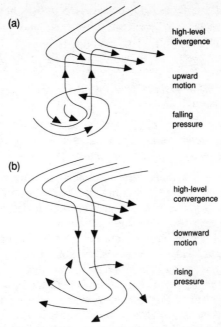

(a)

high-level
divergence

upward
motion

falling
pressure

(b)

high-level
convergence

downward
motion

rising
pressure

Figure 5.3 Upper winds largely dictate the development of weather systems.

(a) Upper winds may take away more air than they bring, leading to ascent of air to replace it and falling surface pressure.
(b) Upper winds can bring more air than they remove, leading to descending (subsiding) air and rising surface pressure.

quantities of latent heat, which adds to the buoyancy of ascent and intensifies the depression.

While upper winds continue to remove high-level air more quickly than it is replaced lower down, surface pressure in the depression will keep falling, and the low-level winds spiralling into it increase. Over a period of days the circulation around a depression extends higher and higher in the atmosphere until it eventually reaches jetstream levels. At that stage, if not before, the jetstream will become distorted and the main flow will meander elsewhere, divergence aloft will cease, ascent stop and the depression start to fill. Its life cycle will be nearly over.

To us the importance of depressions is clear – they bring cloud, wind and rain. They also have important functions in a wider context. The track of most middle latitude depressions in both hemispheres curves polewards to the north-east or south-east, and they are the most important component of the general circulation in redistributing excess heat from tropical regions towards the poles. Not only that, we saw in Chapter 4 that heat enters the atmosphere largely by the sun heating the ground; depressions are an important mechanism in spreading this heat energy through the depth of the troposphere.

We will now look more closely at the two main driving forces behind frontal depressions and how they fit together in the weather machine: these are the jetstreams and weather fronts.

A zone where a jetstream curves marks an area of imbalance, and the changing direction of the wind involves not only rotation as we saw in the previous chapter, but also acceleration or deceleration. Cyclonic curvature (anticlockwise in the northern hemisphere) which occurs in the trough of a wave, represents an increase in spin relative to the Earth and acceleration outwards, causing divergence at that level. This, as we have seen, leads to a fall in surface pressure. Often this pressure fall is temporary, and the curve in the upper flow moves on. Where the pressure fall coincides with a weather front, a wave on that front develops, which may move together with the trough in the upper winds. In that case, the pressure fall becomes concentrated and prolonged, reinforced by the release of latent heat. A fully fledged depression forms, as shown in Figure 5.4, and itself distorts the jetstream further. Once the process is underway, the depression may persist for several days and influence the circulation around the whole hemisphere.

If this was the complete story, we might expect an active depression to be marked by a more or less circular blob of cloud and rain with strong winds blowing anticlockwise around it. Satellite pictures show graphically that this is not so, and so does experience. Areas of cloud tend to take the form of more or less well-defined bands, usually spiralling into the centre of a depression. We will now see why this should be so.

Weather fronts and air masses

There are vast areas of the Earth's surface over which the terrain varies little, and neither does the balance between incoming solar heating and outgoing longwave radiation. The most obvious of these are the oceans, then there are polar regions and the large continents. In these zones, provided the weather is settled over a long period, then the atmosphere through much of the troposphere reaches an equilibrium, and temperature and humidity may vary little over hundreds of thousands of square kilometres. Such a region is called a **source region** and the body of air with more or less uniform characteristics is an **air mass**.

The character of air masses varies greatly. Over continents air mass temperature depends greatly on the time of year, because in winter most continents are very cold, and in summer hot. Oceans, on the other hand, especially in the tropics, vary little in temperature through the year, and maritime winds from these source regions are invariably humid and warm. The main air masses and their characteristics are shown in Table 5.1.

Table 5.1 Air masses

Region of origin		Abbreviation	Season	Characteristics		
Tropical	Continental	Tc	Summer	Hot	Dry	Sunny
			Winter	Cool	Dry	Frosty
Tropical	Maritime	Tm	Summer	Warm	Humid	Wet
			Winter	Mild	Moist	Drizzly
Polar	Continental	Pc	Summer	Warm	Dry	Sunny
			Winter	Very cold	Dry	Misty
Polar	Maritime	Pm		Cool	Moist	Showery
Returning* polar	Maritime	rPm		Mild	Moist	Cloudy
Arctic	Ice/Snow	A	Summer	Cold		Showery
			Winter	Very cold		Snow

Note: *This is air of polar origin which has curved well south over the North Atlantic and approaches the British Isles from the south-west. It is milder, cloudier and less showery than Pm air which takes a more direct track.

The importance of the concept of air masses has diminished in recent years, because of the modern mathematical approach to weather forecasting. Nevertheless, when winds are expected from a particular source region, consideration of the associated air mass, together with likely modifications during its journey, enables a useful first estimate to be made of the likely weather. Boundaries between air masses are often vague and diffuse, but where they are concentrated over a relatively short distance they are termed **weather fronts**, **frontal zones**, or just **fronts**. These are important, because temperature and humidity differences across them often give rise to cloud and rain. When a depression starts to form with the low-level winds beginning to blow cyclonically, then the warmer (and usually moister) air mass tends to be forced above the colder heavier air. Areas of ascent are, therefore, concentrated in the frontal zones, readily producing cloud and rain. We will look at fronts in more detail later.

The boundary between air of polar and tropical origin is often abrupt, and can often be seen on satellite pictures and followed from observations around much of each hemisphere. It is termed the **polar front**. Frequently a procession of deep depressions sweep eastwards across the wide oceans in both hemispheres. Each depression can usually be traced back to a small wave on the polar front. The less frequent but equally sharp boundary between an air mass of Arctic origin and others, is termed the **arctic front**. When this crosses southwards in winter over the United Kingdom, North West Europe or North America it often brings snow, and heralds the beginning of the coldest northerly outbreaks.

We are now in a position to consider in more detail the way a frontal depression develops across the frontal zone between two air masses (see Figure 5.4). We start with a more or less uniform polar front separating air of polar origin and warm air from the tropical seas to the south. The boundary is initially quite diffuse because both air masses are far from their source regions, and will have been modified by the conditions encountered on the way. When a jetstream arrives with its area of strong outflow or divergence at high levels, air begins to ascend over this broad frontal zone, generating a cyclonic circulation in the winds at low levels. Soon the temperature distribution becomes distorted into more concentrated bands, and the warm moist air of tropical origin is drawn into the circulation of the low in a wedge shape, called a **warm sector**.

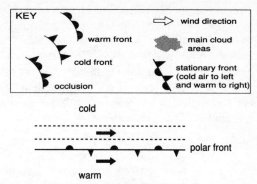

cold

polar front

warm

(a) Polar air faces tropical air across the polar front

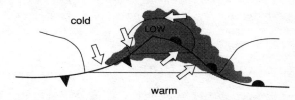

warm

(b) A perturbation in the upper winds starts pressure falling, and surface winds begin to blow anticlockwise, creating a wave on the front.

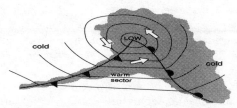

(c) As the depression deepens the fronts develop stronger temperature gradients across them, and become more active. Winds increase. A warm sector develops.

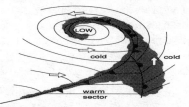

(d) Much later occlusion begins as warm air in the peak of the warm sector becomes lifted above the surface.

Figure 5.4 Formation of a warm sector depression

If the jetstream above moves quickly on, then the immature beginnings of a depression will soon fizzle out, or move quickly along the front as a non-developing **wave** – exactly analogous to a wave on the sea. These immature **frontal waves** can sometimes travel along a front for 2000 km (1250 miles) or more, and produce considerable rain as they sweep through. They pose a particular problem for forecasters because, being small and fast moving, the precise area likely to be affected is uncertain even only an hour or two ahead.

If the jetstream is slow moving it can phase in with, and become distorted by, the forming depression. This increases the divergence in the upper troposphere even more and leads to further deepening. This so-called positive feedback over a period of a day or two leads to a fully fledged and intense **warm sector depression**. In its lifetime of a week or so, it will bring rain and strong winds to many thousands of square kilometres, and impact upon the everyday lives of millions of people.

More about weather fronts

The zones between air masses where the main temperature differences are concentrated in the circulation of the depression, are the fronts. In Figure 5.4 they are shown with the conventional meteorological symbols, which are always placed so that they point in the direction of movement. Thus a **warm front** marks the transition from cold to warm air and a **cold front** the reverse. With time the warm sector in the depression becomes narrower until it is progressively lifted completely above the surface starting at the apex of the wedge. Where this happens the resulting front is said to be **occluded**, and is called an **occlusion**. Even an occlusion may have a marked difference in the character of surface air across it. Once the warm sector starts to occlude the depression usually ceases to deepen. This is because one of its important energy sources – the supply of latent heat from ascending warm moist air – is cut off. As a depression matures and starts to fill, an often fragmented band of clouds can frequently be seen on satellite pictures spiralling into the centre of the system, marking the remains of occluded warm air.

No two fronts are exactly the same, but warm fronts and cold fronts are quite different as we shall see. They are shown schematically in cross section in Figure 5.5. Because ascent of air in a depression is concentrated in the frontal zones, so is much of the cloud and rain. A developing depression usually has active and intense fronts; a mature, filling depression often has weak fronts perhaps giving no rain at all. Furthermore, the weather on a front tends to be more intense the nearer it is to the centre of the depression.

The warm front and warm sector – see Figure 5.5 (a)

At a warm front the warmer air often shelves above the preceeding colder air in a gradual slope for some hundreds of kilometres ahead, as it is forced to ascend. Often the first sign we see from the ground of its approach is the sun (or moon) becoming indistinct behind what seems like frosted glass. This is the effect of thin high cirrostratus ice crystal cloud at a height of around 10 km (6 miles). Sometimes there is an accompanying halo (see Chapters 9 and 10), indeed at night the cloud at first may be so thin that a halo round the moon is the only sign of the cloud's presence. Over the following few hours the cloud thickens, its base lowers, the wind increases, and rain starts. At first the rain is light and fitful, but often nearer the surface front there are

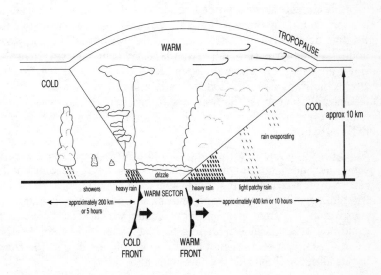

Figure 5.5 (a) Warm front, cold front and warm sector.

bands of quite heavy rain 50 to 100 km (30 to 60 miles) or so across. The surface front usually arrives about 12 hours after the first signs of thickening upper cloud, following a spell of heavier rain. At the front the wind veers, the temperature rises and it will usually become more humid in the tropical air mass following.

In Europe a warm sector brings mild, cloudy weather, but the cloud is usually quite shallow, typically less than 1000 m (3000 ft). It will often give drizzle especially where forced to rise over hills. In summer when the sun is strong, or well south of the low centre, cloud may break to the lee of hills to give spells of warm sunshine. Nights in warm sector conditions are often misty, with more widespread drizzle because the cloud thickens as it cools by longwave radiation from its top, increasing the water droplet concentration (see Chapter 3). Over the United States east of the Rockies and South America east of the Andes, warm sectors are almost invariably dry, warm and almost cloud free. This is entirely due to the Föhn effect, during the long descent of the air down the mountains (see Chapter 10).

The cold front – see Figure 5.5 (a)

Cold air circulating round a depression is heavier for a given volume than warm air. This means that at a cold front the cold air undercuts warmer air ahead of it, contrasting with the warm front. On the other hand, the cold air is less stable, that is to say more buoyant, than the warm air, especially where it passes over warm sea or heated ground. Hence, slowing due to friction is less in cold air, and low-level winds tend to be stronger. At the front itself, therefore, the low-level winds tend to overrun surface warm air to give a cold 'nose', which contributes to blustery winds.

At higher levels the structure of cold fronts varies considerably. In an active and deepening depression the cold front will be continually renewed by ascent and the increasing wind flow. Deep dark clouds will often give heavy rain for an hour or so, and perhaps even hail or thunder as the front passes, with squally winds. There follows a rapid, almost magical brightening as blue skies replace the frontal cloud. Usually, it is not long before deeper cold air behind the front brings ideal conditions for the development of showers.

Sometimes strong upper winds take the warm moist air with most of the thick cloud and rain a long way ahead of the surface cold front.

We may then see only one or two bands of patchy light rain or sometimes nothing at all. This is termed a **split front**, because the weather aloft and near the ground are widely separated, and rain ceases long before the surface temperature shows its characteristic decrease and the wind veers (see Figure 5.5 (b)). Even a weak split cold front, however, may have a sting in the tail, because convergence in the low-level winds at the front, together with deep, cold air sometimes cause heavy showers or even thunderstorms to develop.

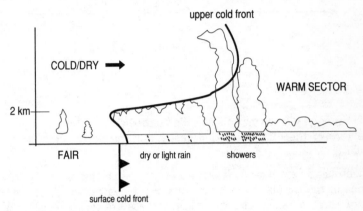

Figure 5.5 (b) Split cold front.

The occlusion – see Figure 5.5 (c)

When warm air in its narrowing warm sector wedge is lifted above the ground the combination of cold and warm fronts is called an occlusion. In its early stages it may behave merely as an extension of an active warm or cold front, depending whether the air ahead of the system or to the rear is coldest, but usually changes of surface wind and temperature are less marked. Occasionally, ascent along the line of an occlusion is sufficient to produce heavy showers or thunderstorms, but usually as a depression occludes it starts to fill, with ascent becoming muted or non-existent. As the occlusion is drawn round the centre of the low, activity on it diminishes until only lines of showers remain. Occlusions, since they are based well above the surface, curve around and not through the centre of their parent depression. Sometimes part of an occlusion curves right round the centre and appears just like a second cold front. This is common over the United Kingdom in winter

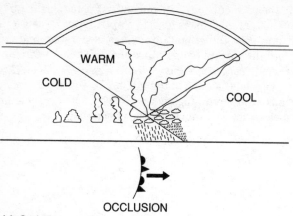

Figure 5.5 (c) Occlusion.

when it is only following this so-called **bent-back occlusion**, well to the rear of the cold front, that really cold northerly winds sweep down.

Sometimes a bent-back occlusion weakens, or a narrow, showery zone forms in the circulation of a vigorous depression rather like a front but with no air mass change across it, these are called **troughs of low pressure** or just troughs.

The conveyor - see Figure 5.5 (d)

No two frontal systems are the same. The preceding discussions and diagrams of Figure 5.5 are two-dimensional simplifications of complex processes. In recent years meteorologists have tried to incorporate the three-dimensional nature of fronts into their simplified models, taking account of the flow along and upwards through fronts towards the centre of depressions. This has become known as a **conveyor** mechanism.

Anticyclones

An **anticyclone** or **high** is an area of high pressure around which surface winds circulate in a clockwise direction (in the northern hemisphere). In many ways it is, not surprisingly, the opposite of a depression or low. The fundamental difference is that, in this case, high-level winds pile up air in the higher troposphere causing surface pressure

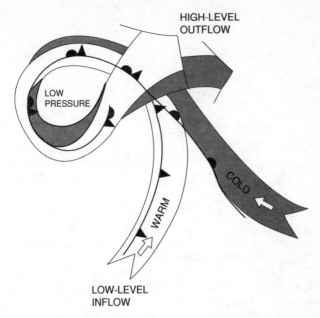

HIGH-LEVEL
OUTFLOW

LOW
PRESSURE

COLD

WARM

LOW-LEVEL
INFLOW

Figure 5.5 (d) Ascent of air in a depression tends to be concentrated ahead of warm and cold fronts. This combined with inflow towards the low centre creates zones where low level air flows into a depression and is gradually lifted before it flows out at high level.

to be higher than elsewhere. As a consequence the air is slowly descending through a large depth of the troposphere, often over a wide area, and winds near the surface take air outwards away from the centre. In other words air is converging at high levels and diverging near the ground. As it descends, air is compressed by higher atmospheric pressure, and warms as the energy from compression is converted to and stored as heat. This is **adiabatic warming**, and is the reverse of the adiabatic cooling of ascending air that we encountered in the discussion of how clouds form in Chapter 3. It means that any pre-existing clouds also slowly warm and evaporate, so that in anticyclones skies become largely cloud-free. That is why areas of high pressure are often associated with fine, sunny weather, with rarely anything much in the way of rain or snow.

Near the surface, air in an anticyclone is forced to flow outwards away from the centre of high pressure. As we saw earlier, the effect of the Earth's rotation then comes into play with the Coriolis force causing

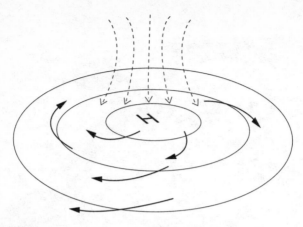

Figure 5.6 (a) An anticyclone is an area of high pressure, with air descending slowly from high levels and warming as it does so because of compression. At the surface winds flow clockwise and outwards across the isobars.

the winds to blow clockwise (in the northern hemisphere), shown schematically in Figure 5.6 (a).

The rate at which air descends is slow, almost invariably less than 1 kn or 0.5 m/s. Nevertheless, anticyclones often persist over many days or even weeks so that the warming and drying of the descending air can be considerable. Often a good deal of the outflow occurs a kilo-metre or so above the surface, with the lowest layer of air remaining relatively cool and perhaps moist. There develops a marked but shal-low transition zone in which there is a rise of temperature with height, or inversion. A typical temperature and humidity profile from the sur-face to the tropopause through an anticyclone is shown in Figure 5.6(b).

Initially this inversion may develop quite high up, perhaps above a pre-existing layer of cloud which took some time to dissolve; or there may be two or three smaller inversions as the weak remains of long-decayed fronts are overcome. Gradually, however, as outflow proceeds, the inversion becomes lower and lower, and over a period of several days it often descends to within 1000 m (3000 ft) of the surface. In very persis-tent anticyclones the inversion may actually reach the surface, but it is much more common for hills and mountains to protrude through

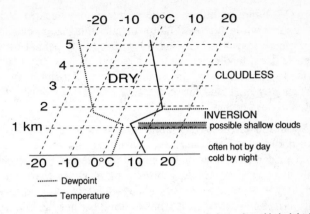

Figure 5.6 (b) The typical variation of temperature and dewpoint with height in an anticyclone. The air becomes warm and dry, and often cloud-free. However, often the greatest warming does not reach down to the surface, and there is a zone where the temperature rises with height, sometimes marked by shallow cloud and trapping pollution beneath it, especially in winter.

into the warmer air above. In this way they may, unusually, be several degrees Celsius warmer than low-lying ground nearby.

The descent and warming of air in anticyclones is called **subsidence**, reflecting the idea of the air above piling up and subsiding. Associated temperature inversions are called **subsidence inversions**, and they are important because they trap pollutants in the layer near the Earth's surface (see Chapter 11). They also explain why anticyclones are not always cloud-free. When the surface air is moist, and either solar heating or turbulence lifts it above its condensation level, cloud forms just below the inversion. Also at night, radiation from the top of the moist layer may cool it enough to form cloud. Often cool moist winds blowing in from the sea during autumn and winter form cloud as they mix with colder air over land, and daytime heat from the weak sun may be insufficient to dissolve it. Since the formation, extent and persistence of such cloud depends on small variations in temperature, humidity and wind, it is difficult to forecast.

Where anticyclones are cloud-free with the usual light winds, they provide ideal conditions for maximum night-time radiation cooling and the formation of frost and fog. These are so called **radiation nights**, well known to gardeners, fruit growers and astronomers!

Where this happens, hill and mountain tops may bask in early morning sunshine while surrounding towns and villages are cloaked in fog or low cloud. In winter, extensive low cloud may persist all day and for many days, leading to what has become known as **anticyclonic gloom**. In summer, however, strong daytime sunshine usually soon disperses any mist and low cloud that has formed during the short night, and days become hot and hazy.

A **ridge of high pressure** is similar to a weak anticyclone. Technically, it does not have a closed isobar on a weather chart, much like a contour around a ridge on a conventional map of topography. From the meteorological point of view, ridges generally move faster than anticyclones which may be stationary for weeks. Consequently, while a ridge almost always brings fine weather, it tends to bring neither the intense heat of summer nor the penetrating cold and gloom of winter.

A **col** is an area where the pressure distribution is flat, lying between troughs and ridges, see Figure 5.7. Weather in cols varies considerably and depends almost entirely on what is happening in the upper air.

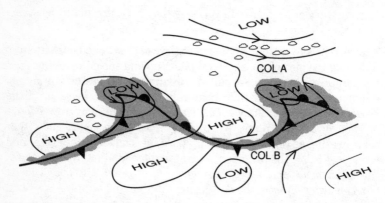

Figure 5.7 A col on a weather map is an area lying between two highs and two lows with little wind. Here there are cols at A and B.

──── Non-frontal depressions ────

In temperate latitudes anticyclones and frontal depressions comprise the main weather features, their intensity depending on latitude, season and the topography of the area concerned. Other weather systems assume greater importance in many parts of the world, especially subtropical and equatorial regions.

Monsoons

While the general circulation of the atmosphere outlined in Chapter 2 drives the weather around the globe, the distribution of oceans, continents and large mountain ranges has a considerable influence. We saw earlier in this chapter how local sea and land breezes are generated in coastal regions by differences in solar heating. **Monsoon winds** are driven by the same type of circulation but by seasonal differences of temperature between continents, Asia in particular, and the tropical oceans. The monsoon climate is immensely important over wide areas of the world and is discussed further in Chapter 12.

Hurricanes or tropical storms

A **hurricane** is an intense tropical depression around which surface winds often reach 100 kn or more. In the Beaufort scale of wind speed described in Chapter 6, hurricane force 12 corresponds to winds exceeding 64 kn. In Japan and the Philippines in the western Pacific hurricanes are called **typhoons**, and in the Indian Ocean **tropical cyclones** or just cyclones, though this is also sometimes confusingly used to refer to the depressions of temperate latitudes.

Hurricanes invariably form over tropical oceans, usually between 5 and 15 degrees latitude. They are usually of relatively small diameter compared with the depressions of middle latitudes – being some 200 to 1000 km (125 to 650 miles) across, and comprise great masses of cloud spiralling into a deep low pressure centre, producing torrential rain along with extremely strong and damaging low-level winds. Strangely, at the centre of a hurricane, over a distance of 10 to 50 km (6 to 30 miles) there is a complete contrast, with well broken cloud and light winds. This 'eye of the storm' can often be seen on satellite pictures and is useful in locating the exact position of the hurricane centre.

On the poleward side of the ITCZ the trade winds curve westwards. As they leave the calming influence of subtropical high pressure, areas of shower clouds form over the warm tropical ocean, often in clusters some 100 to 300 km (60 to 200 miles) across. Most clusters form and decay over a few days, but occasionally one grows explosively and spawns a hurricane; why?

We saw earlier in this chapter the main component that generates a middle latitude depression is an upper wind pattern to provide divergence and ascending air, to draw in cyclonic low level flow. The trigger for a hurricane to develop appears also to be a distortion in the upper wind flow pattern in the high troposphere. This similarly produces outflow aloft which provides the initial impetus to encourage pre-existing cloud to expand. However, the most important factor thereafter appears to be the hot, humid air at low levels and the huge reservoir of heat energy that the tropical oceans provide. It has been found that hurricanes never occur when the sea temperature is less than 27 °C. No fronts are involved; the energy that becomes concentrated in a hurricane is almost entirely provided by latent heat, hidden in the huge volume of hot and humid air sucked up into the initially weak circulation, itself adding massively to the suction power as water vapour condenses into cloud. In low latitudes the Coriolis force is small, and there is little sideways deflection of air being drawn towards the centre of low pressure. Nonetheless, the air does not flow directly to the centre of low pressure in the middle of the hurricane because pre-existing rotation is concentrated about a smaller and smaller radius as it is drawn inwards, generating centrifugal forces much as water flows down a plug hole. Air spirals anticlockwise (cyclonically) towards the centre (northern hemisphere) at increasing speed, and is lifted to form the characteristic 'catherine wheel' cloud shape around the clear 'eye' which it never reaches. Movie sequences of satellite pictures show cold, high cloud streaming outwards, in the diverging upper flow.

Hurricanes are amongst the most damaging natural phenomena on Earth. They cause death and destruction from their severe winds and flooding from rain and tidal waves. One small blessing is that for continued sustenance hurricanes need the underlying warm ocean; once they start to travel inland they lose their fuel (of latent heat) and invariably weaken quickly. The life cycle of most hurricanes is over in a week, although a few last for more than two weeks. Many spend their life at sea proving a severe hazard to shipping; the most damaging either cross islands or move inland before they dissipate. Most hurricanes drift westwards and curve away from the equator, but they can change direction abruptly. Forecasts are improving, but accurate prediction of tracks for more than a day or two ahead is rarely possible. Many over the North Atlantic curve north and then eastwards as they weaken, and then reinforce depressions in middle latitudes. Forecasters need to be aware of this, because such depressions often bring widespread gales and rain to parts of North West Europe.

The hurricane season is generally the late summer and early autumn in each hemisphere when the seas are warmest. In the northern hemisphere from July to October the areas most at risk are those bordering the Caribbean, the Gulf of Mexico, and the South China Sea. In the Bay of Bengal, however, they are most frequent from April to June and from September to December. In the southern hemisphere hurricanes mostly affect islands in the south Indian Ocean west of South Africa, and parts of north and west Australia from December to March.

Throughout the year a hurricane watch is carried out by various meteorological services pooling the latest information. In the early stages, when winds are below gale force 8 on the Beaufort scale (see Chapter 6), a system is called a **tropical depression** and will be allocated a letter and number. Should winds increase (or be believed to have increased) to between gale force 8 to violent storm force 11 it will be termed a **tropical storm** and be given a name. The system only becomes a fully fledged hurricane when hurricane force 12 is believed to have been reached. At the beginning of each hurricane season a list of names is allocated for each ocean. The names on each list have initial letters running alphabetically and alternating male/female. For example, in mid-August 1995 in the Pacific, hurricane Erik had just dissipated west of Mexico with Flossie forming nearby; while in the Atlantic sector Gabrielle was weakening in the Caribbean, but further east a violent (and rather older) Felix was heading for Bermuda. Yet to be born in the Atlantic were Humberto, Iris, Jerry, Karen, Luis and so on.

Polar lows

A **polar low** has two factors in common with a hurricane: it is generated and maintained largely by latent heat energy from a warm sea, and is usually less than 300 km (200 miles) across and often much smaller. The differences, however, are many. Polar lows, as the name implies, form in high latitudes. They are rarely associated with particularly strong winds, neither do they bring heavy rain, but they do often bring snow.

Polar lows form as deep cold Arctic air flows southwards across relatively warm seas. Consequently, they favour areas where warm sea currents push polewards, such as the Gulf Stream around and north of the British Isles. They also develop some of the characteristics of frontal depressions as they mature, when temperature contrasts arise due to cold air warming as it pushes southwards over warmer water, and 'fronts' or troughs develop. It appears that a small fluctuation in the upper wind pattern causes enough high-level divergence to encourage local deep convection and a fall of pressure. This generates the classic low-level circulation of a depression, and latent heat released as the cloud forms keeps the system going.

Polar lows rarely last more than a day or two but they can produce large snowfall and considerable disruption. Because they start as small disturbances, below the resolution of mathematical models (see Chapter 7) and usually originate in areas where observations are sparse, they are difficult to forecast.

Part Two

USING WEATHER INFORMATION

6

OBSERVING AND REPORTING THE WEATHER

To forecast the weather at any one place it is essential to know what is happening elsewhere. Weather systems move and develop. To keep track of them meteorologists have developed a huge network of observations of many different kinds, that span the globe. Together these enable a picture to be assembled, not only of the weather at the Earth's surface, but also through the depth of the troposphere and the lower stratosphere. Of course, we are most interested in how the weather affects *us*, but it is important to remember that the atmosphere is three-dimensional. What happens here tomorrow depends greatly upon what is happening far away in the upper atmosphere today. No matter how sophisticated new techniques become, it remains essential to observe the weather. Indeed, advances over the last few years have made it even more important that weather observations are carried out accurately and regularly worldwide, in the upper air as well as at the surface.

A meteorologist portrays the weather picture as a series of charts, showing the distribution of pressure, wind, cloud or rainfall, for example, at fixed times. At main Forecasting Centres another type of 'picture' is composed, formed of large series of numbers derived from many thousands of observations. These enable the state of the atmosphere to be represented in a computer. Accurate observations not only provide an essential basis for weather forecasting, but over many years enable climates to be assessed and compared. It would be a headstrong individual indeed who planned his or her holiday on a single observation or even a week's observations!

In this chapter we consider how the various meteorological measurements are made, looking at conventional instruments, which have, in some cases, been used for hundreds of years, and also the latest remote sensing techniques. These obtain information from many hundreds of kilometres away, and include radar and satellites.

——— Surface observations ———

The most accurate observations are made by direct measurement using high-quality, correctly calibrated instruments. The instruments should be positioned so that readings represent conditions in the free atmosphere, undistorted by nearby buildings, trees or other obstacles. This ideal is rarely attained, but meteorologists have established standards of exposure for instruments, offering a good approximation. On board ship and on buoys the problem is exacerbated by the moving platform, often well above the surface of the sea, and also by blowing spray.

Below we consider the main elements of an observation in turn.

Temperature

The **temperature** of a substance is a measure of its energy level due to molecular motion. Measurements must be made on a readily understood **temperature scale** based on consistent and easily reproduced reference levels.

The **Celsius** or **Centigrade** (°C) scale takes its zero as the freezing point of water, and 100 degrees as the boiling point of water (they vary with pressure, so the strict specification includes a standard pressure). While this scale is now used globally in meteorology, temperatures are still converted into degrees Fahrenheit for the general public in some English speaking countries. In the **Fahrenheit** (°F) scale the freezing point of water is 32 degrees and the boiling point

---

Temperature conversion between Celsius/Centigrade and Fahrenheit

From °F to °C add 40, multiply by 5, divide by 9 and subtract 40
From °C to °F add 40, multiply by 9, divide by 5 and subtract 40

---

212 degrees. The easiest way to convert temperatures between the two scales uses the fact that they coincide at minus 40 degrees.

The **Kelvin** (°K) or **Absolute** scale of temperature, is zero at –273.15 °C where theory shows that molecular motion should cease. It is never used in meteorological observing but must be used in most scientific calculations. To convert from degrees Absolute to degrees Celsius/Centigrade add 273 (approximately), and vice versa.

Thermometers exploit changes that occur in many substances as the temperature rises or falls. For example, most metals and many liquids expand proportionally as their temperature increases, and this forms the basis of the common **mercury-in-glass thermometer**. This comprises a reservoir of mercury in a glass bulb with an enclosed, uniform and narrow tube leading from it. A rise of temperature causes the volume of mercury in the bulb to increase, and the extra mercury is forced along the narrow tube so that it looks like a silver thread. Conversely, a temperature fall causes the mercury thread to contract and retreat towards the bulb. Every thermometer is calibrated to relate the length of thread to the temperature.

Mercury freezes at –38.9 °C; it would be useless below that temperature in a thermometer. Where very low temperatures are possible alcohol is used, which freezes at –114.4 °C.

To measure true air temperature, sometimes called the shade temperature, a thermometer must be positioned away from direct or reflected sunshine, where the air may circulate freely around it but shielded from rain or snow. Meteorologists have agreed air temperature should be measured at a height of 1.25 m (4 ft) above the ground.

To approach this ideal exposure the **Stevenson screen** has been widely used for many years. Designed by Thomas Stevenson, the father of author Robert Louis Stevenson, this screen is a wooden cupboard mounted securely on legs. It is louvered to allow air to circulate freely through and around it, while keeping rain and leaves out. A double roof and white finish insulate instruments within from direct or indirect heating by the sun. Thermometers in the screen are placed on racks out of contact with the floor or sides so that air can circulate freely round them. The Stevenson screen should always be sited well away from buildings and trees, with its hinged door facing directly away from the midday sun.

Special thermometers to measure maximum and minimum temperatures are usually also sited in the Stevenson screen. At first sight they seem identical to the ordinary thermometer, but they each have important modifications.

The **maximum thermometer** records the highest temperature since it was last reset. It has a constriction in the narrow tube above the bulb of mercury. Rising temperature readily forces expanding mercury through the constriction, but subsequent contraction as temperature falls leaves the upper part of the thread stranded beyond the constriction. Thus the end furthest from the bulb still registers the maximum temperature reached (see Figure 6.1 (a)). It is reset by careful shaking.

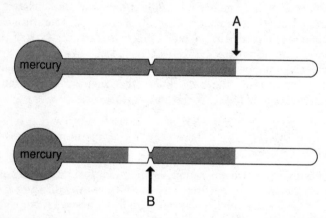

Figure 6.1 (a) Maximum thermometer: temperature reaches a maximum at A, and when it subsequently falls the mercury thread breaks at the constriction B, leaving the reading unchanged.

The **minimum thermometer** records the lowest temperature since last reset. It uses alcohol rather than mercury, and includes a small needle within the thread of alcohol in the thermometer stem. As the alcohol contracts on cooling, the needle is drawn back towards the bulb, but is left behind when the temperature subsequently rises. Consequently the end of the needle furthest from the bulb points to the minimum temperature reached (see Figure 6.1 (b)). This thermometer

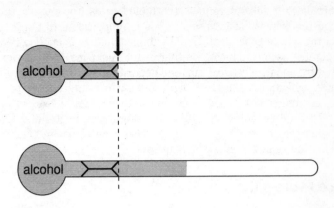

Figure 6.1 (b) Minimum thermometer: the temperature falls to a minimum at C, taking the needle down as it does so. When it rises subsequently the needle remains.

is reset by tilting it sufficiently for the needle to slide back to the end of the alcohol thread.

A **thermograph** provides a continuous record of air temperature on a special chart called a **thermogram**. It comprises a pen arm attached to one end of a metal coil with the other end fixed. But this is no ordinary coil, it is made of two different metals bonded together (sometimes called a bi-metallic strip). Because different metals expand (contract) at different rates with a rise (fall) of temperature, this coil unwinds on heating and tightens on cooling. Such variations tilt the pen arm and the nib records the corresponding temperature as a continuous trace on the suitably calibrated thermogram mounted on a slowly rotating drum.

It is important to measure and predict temperatures at ground level and below, especially for horticulture and transport industries which are affected by frost. Many meteorological stations measure night minimum temperatures at ground level, over both grass and concrete, known unsurprisingly as **grass minimum** and **concrete minimum** temperatures. In the United Kingdom selected stations also report **earth temperatures** at depths of 5, 10, 20, 30 and 100 cm (2, 4, 8, 12 and 40 inches).

It is prohibitively expensive to station observers at every location from which observations would be useful, and systems have been designed to measure and transmit readings automatically. None of the traditional thermometers so far described is suitable for this. Fortunately, the electrical conductivity of many materials is proportional to temperature, and this property is ideal. Usually a fixed voltage is applied to a high resistance semi-conductor or thermistor requiring little power. The resulting small electrical current depends upon the electrical resistance, which varies in turn with the surrounding air temperature. The current may be used to produce either a digital readout of temperature, or a reading on a suitably calibrated ammeter dial. Often it is arranged to generate a radio or telephone signal in combination with other remote reading instruments.

Pressure

Atmospheric pressure is the weight of atmosphere on a fixed unit area (e.g. a square centimetre or square inch), but unlike a solid object, pressure in a fluid acts not just downwards but in all directions. Historically atmospheric pressure was measured in centimetres or inches of mercury, this being simply the height of a column of mercury with the same weight per unit area, about 750 mm (29 inches). Meteorologists now use **millibars** or mb for short. At the Earth's surface the atmospheric pressure varies from around 950 mb, at the centre of a deep depression, to 1050 mb in an intense anticyclone. At the tropopause the pressure is usually between 400 mb and 200 mb; at the outer reaches of the atmosphere pressure is near zero (see Chapter 2, Figure 2.1).

A **millibar** is one thousandth of a bar, which is equal to one million dynes per cm^2 or one hundred thousand newtons per m^2. A dyne is the force needed to accelerate 1 g through 1 cm per second per second, and a newton is the force needed to accelerate 1 kg through 1 m per second per second. A millibar is exactly the same as a **hectopascal** or Hp which is now sometimes favoured.

Heights in the atmosphere are often referred to by meteorologists in terms of pressure, and upper air charts usually relate to fixed pressure levels. On these charts contours reflect the varying height of the particular pressure level, exactly analogous to contours of topography on an ordinary map. This is done because contour spacing relates directly to the geostrophic wind, regardless of the level.

Atmospheric pressure is easy to measure accurately and, because variations relate closely to wind and weather, it is the most important element of meteorological observations.

Surface weather charts used by meteorologists almost invariably include **isobars**, lines joining points with the same mean sea level (**MSL**) atmospheric pressure. The continuity of isobars together with the skill of an analyst, not only show areas of low and high pressure and the position of fronts, but enable the direction and strength of the wind to be deduced where there are no observations. Their orientation shows the direction (Buys Ballot's law) while the spacing and curvature reflect the speed at any latitude. Weather charts are discussed further in Chapter 7.

A **barometer** measures atmospheric pressure. There are two main types: mercury and aneroid (meaning 'without liquid') barometers.

The **mercury-in-glass barometer** in its simplest form is an inverted glass tube closed at the top, standing in a reservoir of mercury open to the atmosphere (see Figure 6.2(a)). Since the mercury in the tube has no air above it, the weight of the column must be entirely supported by the pressure of air on the reservoir. The atmospheric pressure can be read directly from the suitably calibrated mercury column, but for accurate measurement corrections must be applied. If the instrument is warmer (colder) than its calibration temperature then mercury and tube will

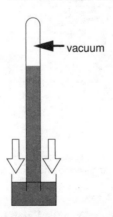

vacuum

Figure 6.2 (a) Mercury-in-glass: the weight of air, or pressure, on the reservoir of mercury is balanced by the column of mercury which has no air above it. Pressure is measured by the length of the column.

have expanded (contracted), and the reading will need adjusting. An additional small and constant correction might be needed if the local force of gravity differs from that assumed in the calibration. More importantly if, as is normally the case, the MSL pressure is required, then a factor will need adding to correct for the difference between the height of the instrument and mean sea level. Mercury barometers are accurate, but expensive and fragile, with the added disadvantage of using mercury which is highly poisonous.

Aneroid barometers are now almost universally used, being smaller, cheaper and more robust than the mercury-in-glass type. The simplest of these comprises a sealed container made in the form of 'bellows' which is a partial vacuum, able to easily expand and contract along one axis (see Figure 6.2(b)). Any increase (decrease) of atmospheric pressure causes contraction (expansion) of this container, and this is magnified by a system of levers connected to a pointer and a scale. Alternatively, a simple electrical system is used to enable readings to be taken remotely. The latest **precision aneroid barometers** use a light pivoted and counter balanced electrical contact arm, which is manually adjusted by the observer to make contact, reducing friction to negligible levels.

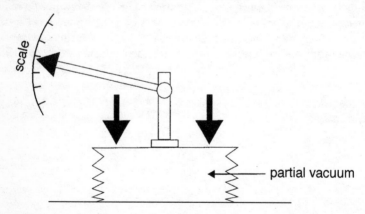

Figure 6.2 (b) Aneroid: changes in pressure cause the aneroid capsule to expand or contract, and readings are registered either by an electrical system or simple pen arm.

Because the aneroid barometer is easily attached to a system of levers and a pen, it forms the basis of the **barograph**, which records variations of atmospheric pressure with time on a chart, or **barogram**, fixed to a rotating drum. A barogram often records pressure variations over a week and sometimes a month, and enables an observer to see recent changes and trends. The change of pressure, or **pressure tendency**, over the three hours preceding an observation forms part of the standard surface **synoptic** message. (In meterology 'synoptic' is used widely as a substitute for 'weather' in describing charts, observations and stations.) Frictional effects make barograph accuracy much inferior to the standard barometer of whatever type.

An aneroid barometer forms the basis of the pressure altimeter, commonly used in aircraft until the advent of radio altimeters and is touched upon in Chapter 2. It is also the basis of almost all domestic barometers, although magnificent antique mercury-in-glass instruments can still be found.

Wind and gusts

Wind is the horizontal velocity of air at a point, normally specified as the direction relative to true north from which the air is flowing, and its speed relative to the surface of the Earth. Wind speed, is by international agreement, measured in knots though some nations favour metres per second (m/s). Wind direction is invariably measured in degrees from true north. If a wind direction changes in a clockwise sense, for example from south to south-west, then it is said to **veer**, to change anticlockwise is to **back**.

A knot is 1.152 mph, or 0.514 m/s. The origin of this perverse unit, the **knot**, is rooted in naval history. A knot is one nautical mile per hour, where a nautical mile is one minute (one sixtieth of a degree) of latitude at the Earth's surface; 8 kn are roughly equal to 10 mph and 4 m/s.

In meteorological observations the **mean wind** is usually an average taken over ten minutes preceding the time of observation. A **gust** is a fluctuation above the mean typically lasting from a few seconds to a few minutes. Windflow is always turbulent due to the roughness of the Earth's surface. Gustiness is much increased by the presence of buildings, trees and rocky or hilly country, but at the same time the mean wind is decreased by friction. For this reason winds over the sea are usually less gusty than over land, and this is noticeable in a low- flying aircraft on crossing a coast.

We usually measure and experience gusts at a point, but they are actually eddies in the flow measuring tens or hundreds of metres across, which can be seen travelling through trees or across corn fields. Each gust is but a small part of a process spanning the whole spectrum of motion in the atmosphere. Very large-scale features of the general circulation generate jetstreams and large weather systems, which in turn cascade their energy into smaller-scale winds associated with fronts and individual clouds. Near the surface under the influence of convection and friction, these winds are broken down into even smaller eddies, until eventually the process reaches the molecular scale, where the energy of motion is finally converted into minuscule heating.

To avoid the worst distortions from local eddies and turbulence, meteorologists specify that the surface wind should be a measurement of the undisturbed flow 10m (30 ft) above the surface of the ground over a flat area well away from trees or buildings. In practice readings are likely to be seriously distorted only if taken nearer to any obstacle than about three times its height. Even if this condition cannot be fulfilled when the wind is from a particular direction, it is usually possible to derive a correction factor by comparing readings over a long period with the nearest well-exposed instrument.

Anemometers measure wind speed and direction, utilising the effect of the wind itself. The simplest and most common is the **cup anemometer**, which comprises three or four lightweight hemispherical cups mounted on a freely rotating vertical spindle to measure speed, and a vane to measure direction. The cups are equally spaced in a horizontal circle with their openings all facing clockwise or anticlockwise. The rate of rotation is then proportional to the wind speed. The **wind vane** is simply a horizontal arm mounted on a vertical spindle with a vertical flat plate or fin lying along the arm at one end and a pointer at the other. The force of the wind turns the vane until the fin lies in the direction towards which the wind is blowing, and the pointer then gives the wind direction. Unlike the traditional weathercock on top of old buildings which is read directly, all professional anemometers convert direction readings into electrical signals which are transmitted to remote dials or digitally to receiving stations.

Rotating cups must be light to respond to light winds, but also robust to survive the strongest gusts, together with hail, snow and icing.

Normally a remote wind speed read-out is obtained by connecting the spindle to a small electrical generator, and transmitting the signal to wherever it is required. On large airfields dials from a runway anemometer are often duplicated at locations miles apart.

Electrical signals from a cup anemometer may be channelled to an **electrical anemograph**, which produces a continuous record of wind speed and direction either digitally for computer processing or from two pens side by side on a suitably calibrated chart or **anemogram** (see Figure 6.3).

The **pressure-tube anemograph** is no longer used operationally. It comprised simply a hollow tube held by a vane with its one open end pointing into the wind. The pressure exerted in the tube depended on the wind speed and was used to operate a system of pen arms to give a continuous record. The same principle is used in the Pitot tube which is the basis of the simplest aircraft air speed indicator.

A **hot wire anemometer** measures the wind speed by its cooling power as it blows across a fine heated wire, the electrical resistance of which varies with temperature. This instrument is extremely sensitive, and useful for accurate measurement of small-scale wind fluctuations in the laboratory, but it is rarely if ever used in operational meteorology.

Wind chill equivalent temperatures

The perception of temperature outdoors depends greatly on wind speed. Weather forecasts in cold weather often quote the wind chill equivalent temperature which takes into account extra cooling of the human body due to the wind. There is no accurate method of measuring this – it depends on the amount of bare flesh exposed, clothing worn, humidity of the air, sunshine and amount of shelter from the wind.

At low temperatures, a rough guide is to take the actual air temperature and subtract 1 °C and also the wind speed in m/s.

For example, an air temperature of 0 °C with a wind speed of 5 m/s (10 kn) is approximately equivalent to –6 °C in calm conditions.

The Beaufort scale of wind force

Force	Description	On land	On sea	Speed (kn)
0	Calm	Smoke rises vertically.	Sea like a mirror.	0
1	Light air	Smoke drifts. Wind vane still.	Small ripples.	1–3
2	Light breeze	Wind felt on face. Leaves rustle; vane moves.	Small wavelets, glassy crests.	4–6
3	Gentle breeze	Leaves and small twigs in constant motion. Light flag extended.	Large wavelets. Crests break. Glassy foam.	7–10
4	Moderate breeze	Raises dust; moves small branches.	Small waves but fairly frequent 'white horses'.	11–16
5	Fresh breeze	Small trees in leaf begin to sway.	Moderate waves, more pronounced long form. Many 'white horses'.	17–21
6	Strong breeze	Large branches in motion.	Large waves. Extensive foam crests. Probably some spray.	22–27
7	Near gale	Whole trees in motion.	Sea heaps up. White foam streaks along wind.	28–33
8	Gale	Breaks twigs off trees. Impedes progress.	Moderately high waves of greater length; edges of crests begin to break into spindrift. Foam blown in well marked streaks along wind.	34–40
9	Severe gale	Slight structural damage possible. Tree branches may break.	High waves. Dense foam streaks along wind. Wave crests begin to topple and roll over. Spray may affect visibility.	41–47
10	Storm	Seldom experienced inland. Trees uprooted. Considerable structural damage.	Very high waves with long overhanging crests. Dense large patches of foam blown along wind. Sea surface white. Visibility reduced.	48–55
11	Violent storm	Rarely experienced. Widespread damage.	Exceptionally high waves. Sea covered with long white patches of foam along wind. Everywhere edges of wave crests blown into froth. Visibility much reduced.	56–63
12	Hurricane	Widespread severe damage.	Air filled with foam and spray. Sea completely white with driving spray. Visibility seriously affected.	64 plus

The Beaufort scale of wind force

There are many circumstances where it is impossible to measure the wind speed precisely, and yet studying its effects on the surroundings enables a useful estimate to be made. Nowhere is this more evident than at sea, and in 1805 Captain Beaufort devised a simple numerical scale of wind forces in terms of the resulting state of the sea. The **Beaufort wind scale** is still widely used in area forecasts for shipping, and has been extended for use on land.

Humidity

Humidity is the measure of how much water is held in the air in the form of water vapour gas (see Chapter 3). We can often get a feel for humidity in cold weather by whether or not our breath forms clouds of droplets as we exhale. What happens is that our warm breath mixes with the surrounding cold air and is soon cooled below its dewpoint, with the result that moisture in it starts to condense to form cloud. In hot weather we know it is humid when we feel uncomfortable and sticky. This is because one of the main temperature regulation mechanisms of humans and most animals is the evaporation of perspiration to keep us cool. If the surrounding air is nearly saturated little perspiration can evaporate and discomfort results.

An instrument for measuring humidity is called a **hygrometer** or **psychrometer**. The most common method for many years has been by the use of a **wet and dry bulb psychrometer**. This device sounds complicated, but it could hardly be more straightforward, because it simply comprises two ordinary thermometers. The first is the standard thermometer located in the Stevenson screen, called the **dry bulb** for reasons which will become clear. The second is also mounted in the screen, and is called the **wet bulb** because its bulb is contained in a muslin bag kept moist by means of a wick from a small container of distilled water. The wet bulb records a lower temperature than the dry bulb because it is cooled by the evaporation (see Chapter 3) of water from the muslin into air flowing past, abstracting the heat required from the bulb of the thermometer. The degree of cooling depends on the humidity of the surrounding air.

If the air is very humid, there will be little evaporation from the muslin and little consequential cooling; the difference between the two thermometer readings (the **wet bulb depression**) will be small.

If the air is completely saturated no water can evaporate and there will be no difference in temperature. If, on the other hand, the air is dry, evaporation from the muslin will be rapid with considerable cooling of the wet bulb, resulting in a large wet bulb depression. From measurements of the dry and wet bulb temperatures an observer determines the dew point and relative humidity from tables, a special slide rule or computer program.

Freezing conditions present a problem, first because frozen muslin partly insulates the 'wet' bulb from passing air, and second because once the muslin dries no water will be drawn up through the wick. This is overcome by painting the uncovered 'wet' bulb carefully with distilled water, and noting the lowest temperature while the thin layer of ice remains on the bulb. It is not accurate.

The **whirling psychrometer** comprises wet and dry bulb thermometers mounted side by side on a small frame with a handle about which it is free to rotate. It is whirled rapidly by hand to provide the necessary ventilation, and calculation of humidity proceeds as before. This instrument enables measurements to be taken anywhere, and is commonly used on board ship.

A **hygrograph** provides a record of humidity on a chart, or **hygrogram**. The most common such instrument is the **hair hygrograph**, which makes use of the strange property of human hair to elongate in a humid atmosphere and shrink in dry surroundings. The instrument comprises a small bundle of hairs, fixed at one end, with the other connected through a system of levers to a pen arm. Slight expansion or contraction of the hair bundle is exaggerated at the pen arm, which is calibrated to record relative humidity continuously on a chart mounted around a rotating drum.

Modern psychrometers use more stable chemical and electrical properties, often combined. Many chemical compounds (e.g. common salt or silica gel) are **hygroscopic**, that is to say they absorb moisture from the air; indeed salt sold for table use has small quantities of other chemicals added to counteract this. Furthermore, the electrical resistance of a thin film containing hygroscopic chemicals has been found to vary with atmospheric humidity. This means that the electrical current generated by a fixed voltage across such a film changes with humidity, and can be used to power a suitably calibrated ammeter to give direct readings of humidity, or produce a radio or telegraph signal to a remote observer. These instruments are known as **hygristors**.

When air is cooled it eventually reaches its dewpoint temperature at which it is saturated; further cooling causes dew or frost to be deposited. This dewpoint or frostpoint temperature can be used to deduce the air's humidity at its original temperature. The **frost point hygrometer** is particularly useful in measuring humidity from high-flying research aircraft, where temperatures are well below freezing. In this instrument air is progressively cooled as it passes over a polished metal 'thimble' from which light is reflected to a photoelectric cell. Once frost is deposited more light is scattered, the photoelectric cell responds and this **frost-point** is recorded, from which the humidity of the air can easily be calculated.

Rainfall

Rainfall is defined simply as the depth of water falling over a defined period, or water equivalent in the case of snow and hail. The time over which it is totalled varies depending on the purpose for which the reading is to be used. A climatologist is interested in annual and seasonal values; a river engineer or farmer may be interested in rain over a week or day, or even from a particular storm. As individuals we can be severely affected by rain over a few minutes, especially if caught outside without a coat! Measurements were traditionally made once a day, but modern digital instruments respond at the time of rainfall and can profile individual showers or provide hourly totals as required.

Rainfall is measured by **rain gauge**, the simplest form of which is any receptacle with a known horizontal surface area exposed in the open so that precipitation will fall into it. The volume of water collected, divided by the area of the opening, gives the depth of rainfall since it was last emptied. This is the basis of most rain gauges, but with added features to improve accuracy and to enable readings to be automated.

Evaporation is a problem especially in hot weather and where the instrument is read infrequently. It is minimised by channelling collected water through a funnel into a narrow-necked bottle, where air cannot easily circulate. Collected water is measured in a narrow calibrated glass cylinder where even small amounts can be read accurately.

Each rain gauge should be sited carefully in the open, well away from the shelter of buildings or vegetation. It should not be adjacent to hard surfaces from which splashes might enter, and must itself have vertical faces inside the opening sufficiently deep to prevent rain loss by splashing out. Less obviously, it has been found that a gauge in an exposed location can itself cause eddies which distort the rain pattern over it, causing it to misread. This may be prevented either by sinking the gauge into a shallow hollow, or by building a small turf wall around it.

For snow and hail to be measured accurately the receiving funnel must be heated to melt them, but clearly this heating must not be sufficient to cause undue evaporation. When all is said and done, rainfall varies considerably over quite short distances, especially in showery situations, and there is little point going to extreme lengths to achieve great accuracy.

A **recording rain gauge** is a version of the simple rain gauge adapted to produce a continuous rainfall record either on a paper chart or digitally. This enables both the rate of rainfall at any time and its duration to be deduced. The **tilting siphon** gauge has a float connected by levers to a pen arm which records changes in level on a chart or **hyetogram**, mounted on a rotating drum. When the container is nearly full it automatically siphons. The most common instrument is the **tipping bucket rain gauge**. This has a tilting mechanism with two attached chambers. One or other chamber rests under the receptor funnel until it has collected a small amount of rain, typically 0.2 mm (0.008 inches); it then overbalances emptying itself and bringing the empty chamber immediately under the funnel. The process is repeated every 0.2 mm of rain and the timed series of tips, each producing an electrical pulse, provides a measure of rainfall amount making intensity and duration simple to compute. A network of tipping bucket gauges connected to a central location, can provide an immediate record of rainfall over a wide area. This immediacy is important because such networks are used to calibrate the most exciting and valuable rainfall measuring method of recent years – rainfall radar, discussed later under remote sensing.

Sunshine

The duration of sunshine is important in climatology because of its influence on temperature and the growing season. On a day-to-day timescale it is important to holidaymakers, gas and electricity utilities and indeed to our general sense of well-being. In common with many facets of meteorology, the simple concept of sunshine is difficult to define absolutely; there are problems with twilight and also with thin cloud through which the sun is dimly visible. Sunshine duration is an imperfect indicator of cloud cover: for example, on a winter's day with the sun low in the sky, a half-cover of cumulus cloud will blot out most sunshine, whereas in summer frequent sunny intervals would be expected.

An instrument to measure direct solar radiation is called a **pyrheliometer**. In Britain sunshine was traditionally measured by the **Campbell-Stokes sunshine recorder**, which is still used at climatological stations. This basically comprises a glass sphere located in the open, securely clamped in a frame. The sphere focuses rays from the sun on to narrow charts, made of special card which chars in the concentrated heat, rather than bursts into flame! As the sun 'moves' through the sky, the length of the charred line gives the duration of sunshine. Of course, on many occasions the line is broken as cloud passes across, so that the total is the sum of several discrete burns and requires careful hand analysis. The instrument is simple, robust and long lasting; but it is expensive, labour intensive and not amenable to automation. In addition, the glass spheres prove attractive to would-be fortune tellers! Versions have been made using water-filled flasks in place of the sphere.

Modern instruments register sunshine duration using the temperature rise induced by the sun, either directly by heating a calibrated disc, or indirectly by generating an electrical current from a thermocouple or one or more solar cells. The result may be recorded remotely, with statistics compiled automatically by computer.

Visibility

The simplest definition of visibility – 'how far an observer can see' – is open to many objections. Can see what? A grain of sand? A mountain? How good is his eyesight? Is it night time? Readings of visibility are required to provide measures of the amount of solid and liquid particles, suspended in or falling through the air. Neither the lack of daylight nor the medical condition of the observer can have a bearing on it.

Visibility is pragmatically defined as the greatest distance at which an object of specified characteristics can be seen and identified with the unaided eye, or could be seen and identified if the general illumination were raised to daylight level. A meteorological observer is expected as far as possible to report the minimum visibility over the whole 360-degree sector from his or her station.

During daylight observers assess visibility using a selection of objects in their field of view at various known distances. The objects must be large enough, relative to their distance away, to be easily seen on a clear day: for example mountain peaks, hills and steeples at long range, and buildings, trees and lamp posts over shorter distances. At night estimates can be made by viewing lights of consistent intensity at known distances. Accurate visibility measurements are best made by 'viewing' standard lights using a photoelectric cell shielded from other light. When the photoelectric cell is connected through an ammeter and plotter a continuous record of visibility, in that one direction, can be obtained. This is the simplest form of **transmissometer** or **visibility meter**. With highly consistent lights, a carefully calibrated instrument with stable and accurate photoelectric cells can give excellent results over the lower range of visibility up to 5 km (3 miles) or so, which is the most important. However, when there are directional variations, and this is often the case especially at coastal stations, automatic instruments have obvious limitations. Aviators face another problem; sometimes thin low cloud or haze hardly affects the observed surface visibility, and a pilot can see the ground clearly from directly above, but on his approach down the glide path to the runway the **slant visibility** may be poor.

Visibility is easier to estimate when it is poor and most important. While it is revealing to a forecaster and often important for his customers, it is not a fundamental variable as far as weather forecasting is concerned. This is fortunate because it varies considerably in space and time, and is most susceptible to the influences of civilisation. Visibility changes are sometimes useful to a forecaster as an indication of the onset and intensity of drizzle, rain and especially snow. This is particularly so at remote automatic stations where visibility is measured but precipitation is not.

Present weather

International standards and codes for weather reports are laid down by the World Meteorological Organization (WMO) based in Geneva, and weather data are exchanged all the time across the globe between all nations. Indeed, it is a source of pride amongst meteorologists that political boundaries, even in the depths of the cold war, have hardly existed in their science, though this no doubt partly reflects the universal high level of self-interest. Weather recognises no national boundaries, and anything other than short-range local forecasting requires data from far afield.

To code, transmit and receive information which is essentially numerical, such as temperature, wind and pressure is relatively simple. With more descriptive weather occurrences – fog, rain, snow and such like – international agreement extends to numerical codes for the various alternatives and strict definitions. The standard international code for present weather caters for 100 alternatives, dealing with weather occurring at the time of observation and its intensity, or what has happened in the past hour. For example: if rain is falling, is it light or heavy, intermittent or continuous? If fog is present, has it thickened during the past hour and is it deep enough to obscure the sky? International agreement ensures that these codes can be interpreted all over the world. Furthermore, a chart plotted in China in accordance with WMO codes, is equally intelligible to meteorologists in New York or London. Codes and plotting models are beyond the scope of this book, but definitions and explanations of the main items of present weather are given below.

Rain and showers are distinguished not simply by duration, although a shower would normally be expected to last for less than 30 minutes, but by the way they form. Showers (and thunderstorms) fall from convective clouds (cumulus or cumulonimbus) whereas rain falls from layered cloud (often nimbostratus) – see Chapter 3. Thunderstorms are discussed further in Chapter 10.

Heavy rain falls at a rate of 4.0 mm per hour (0.2 inches per hour) or more; light rain at less than 0.5 mm per hour (0.02 inches per hour) and moderate rain between the two. Heavy showers fall at a rate of 10 mm per hour (0.4 inches per hour) or more; light showers at less than 2 mm per hour (0.08 inches per hour) and moderate showers between the two.

Rain is reported as continuous when it has fallen for the preceding hour with breaks not exceeding 10 minutes; it is otherwise reported as intermittent.

Drizzle comprises drops less than 0.5 mm (0.02 inches) in diameter, and falls from layered cloud (usually stratus). Light drizzle gives negligible run off. Heavy drizzle falls at a rate of 1.0 mm (0.04 inches) or more per hour reducing visibility to 200 m (220 yds) or less, and is rare. Moderate drizzle lies between these.

Freezing rain and **freezing drizzle** occur when the drops either fall through air that is below freezing or, more usually, on to frozen surfaces creating a coating of ice. Both can lead to heavy build up of ice on trees, masts and telegraph wires, and have occasionally disrupted communications over wide areas. Roads and pavements become icy and treacherous rapidly at the onset, and light aircraft may be seriously affected. Neither phenomenon is common.

Freezing drizzle is the most frequent. It usually occurs in long-lived continental winter anticyclones where the ground and lower layers are very cold. Cloud thickens under the subsidence inversion as cooling from outgoing longwave radiation more than outweighs incoming heat from the weak winter sun, and freezing drizzle may become prolonged.

Freezing rain mostly falls ahead of weak winter warm fronts moving slowly over frozen ground and cold low-level air. It starts (like most rain) as snow which melts while falling through a warmer layer some way above the surface. Once freezing rain begins, the rapid build-up of ice creates slippery conditions for a short period, until the release of latent heat and warmer winds melt it. Because it depends largely on local conditions, freezing rain is extremely difficult to forecast.

Snow is precipitation in the form of ice crystals. It usually falls as snowflakes each comprising many crystals stuck together, but in very cold conditions may reach the surface as individual hexagonal crystals, looking rather like sparkling dust.

Heavy snow accumulates to a level depth of greater than 4.0 cm/hour (1.5 inches/hr), moderate snow 0.5 to 4.0 cm/hr (0.2 to 1.5 inches/hr) and slight snow at less than 0.5 cm/hr (0.2 inches/hr).

Snow readily builds into smoothly curved and often deep drifts when blown by the wind against and around obstacles. When the winds are strong, drifting is serious when heavy snow is falling; but dry and powdery lying snow will also drift in strong winds making clearance of roads and railways impossible. In blowing snow, visibility is severely affected. Wet snow, which occurs when the temperature is near freezing, readily compacts and, although it will drift as it falls, will not usually be lifted subsequently by the wind.

A **blizzard** is defined in most countries as a combination of winds of Beaufort force 7 or 8 with moderate or heavy snow reducing the visibility to 200 m (220 yds) or less (United States – 150 m (160 yds) or less and temperature below –7 °C).

A **severe blizzard** involves moderate or heavy snow with drifting, in winds of Beaufort force 9 or more, reducing the visibility to near zero (United States – also a temperature of less than –12 °C). At these extremes, conditions reach '**white-out**', when sky and land merge into complete whiteness, all visual reference is lost and human survival itself is seriously threatened.

Sleet in most areas of the world describes a combination of rain (or drizzle) and snow; in the United States the term is applied to snow pellets or ice pellets (see below).

Snow grains are small opaque pellets resulting from drizzle drops freezing in or below the parent cloud.

Hail takes the form of solid pieces of ice, almost always rounded in shape while rarely, if ever, being exact spheres. Less commonly they take the form of irregular lumps where two or more hailstones have frozen together. Hail grows from collisions with water drops and ice crystals, and requires strong upward air currents for support during its growth. Because of this, it is invariably associated with cumulonimbus cloud, and often coincides with thunder and lightning. It is discussed further in Chapter 3. Small immature hailstones are sometimes called **ice pellets**.

Soft hail, also known as **snow pellets** or **graupel**, is also associated with convective cloud. It comprises opaque pellets of ice, from 1 mm to about 5 mm (0.04 to 0.2 inches) in diameter which, unlike true hail, can be crushed easily. It forms when supercooled droplets collide and instantly freeze, trapping air bubbles between them. This is most common in winter showers when the whole depth of cloud is below freezing and the water content limited.

Fog and **mist** are conditions in which visibility is reduced due to the suspension of water droplets (or ice crystals) in the air. Both may be regarded as equivalent to cloud based at ground level, but most mist and fog form in moist air cooled by contact with the cold ground. Fog is when the visibility is reduced to 200 m (220 yds), below which it affects road traffic. For the purposes of aviation (and in coded observations) 1 km (1100 yds) is the upper threshold. Mist is reported in conditions of reduced visibility, but above 1 km, with a relative humidity above about 95 per cent. In other words the reduction is ascribed to water droplets. In drier air **haze** is reported when visibility is reduced, but in dry air with the visibility below fog limits it would be expected that a **dust storm**, **sand storm** or sometimes **smoke haze** would be appropriate.

Radiation fog is by far the most common type of fog over land. It forms at night when, under clear or nearly clear skies, the Earth's surface cools rapidly by radiation and light winds mix the cooling through only a shallow depth. The cold surface cools the air next to it below its dewpoint and, provided winds are not flat calm, mist and then fog forms. Radiation fog is most common and persistent on clear winter nights when cooling is most prolonged. It is often patchy in nature, reflecting variable rates of cooling and humidity, usually forming first in rural valleys where the coldest air tends to collect.

Hill fog is produced when moist winds are forced to ascend over hills higher than their condensation level, and cloud forms over the upper slopes. Alternatively, pre-existing cloud is often blown on to or across hills or mountains projecting above its base.

Advection fog forms when comparatively warm moist air blows over cold surfaces, and is cooled below its dewpoint. It is much more common over sea than over land for the simple reason that land surfaces are soon warmed by the wind, whereas the cold sea is less responsive. Nevertheless, sometimes behind a warm front, especially over lying snow, advection fog forms widely.

Arctic sea smoke is fog formed when cold air drifts over much warmer water. In this case evaporation from the (warm) water surface proceeds more rapidly than the air in contact with it can absorb it. The excess water vapour condenses out as smoke-like wisps which evaporate as they rise further up through the drier air. This is exactly analogous to the 'steam' which sometimes rises from road surfaces when sunshine follows a shower. It is by no means peculiar to the

Arctic despite its name, but can often be seen on crisp winter nights over rivers or estuaries.

Freezing fog is simply fog at a temperature below freezing. It will normally remain as water droplets unless the temperature is very low and the fog long-lived, when ice crystals may form. However, the droplets freeze readily on impact with cold surfaces producing a build up of rime ice (see below), and making roads and pavements slippery.

Frost (strictly **air frost**) occurs when the air temperature falls below freezing (0 °C or 32 °F). A **ground frost** is when the temperature at ground level is at or below freezing, as measured by a grass minimum or concrete minimum thermometer.

Dew and **hoar frost** are deposited on surfaces on clear, calm nights, when cooling is concentrated in the air next to the cold ground and the temperature falls to below its dewpoint. If the temperature is above freezing then dew drops are deposited. Dew can be confused with droplets exuded by certain vegetation by night, called guttation. When the temperature is below freezing, small white ice crystals form as moisture condenses directly out (sublimates) as ice. Sometimes dew forms before the temperature falls below freezing, and larger frozen dewdrops can be seen amongst the hoar frost. Often following calm clear nights early rays from the rising sun stir the cold air near the surface, and where only dew had formed soon there are mist or fog patches.

Rime (or rime ice) is a build-up of ice which occurs when supercooled droplets are blown against objects at a temperature below freezing. It is white and opaque since air is trapped in it, and often occurs in cloud on mountain tops, where it can build up to several centimetres on the windward side of buildings and instruments. Aircraft icing in cloud is usually formed in the same way.

– Cloud categories and observation –

Few can fail to feel uplifted by white blossoming turrets of cloud, etched against a blue sky. And perhaps just as many depressed when glowering grey cloud seems low enough to touch. The form and texture of cloud, affected as they are by height, density and illumination seem boundless. In fact, there only two fundamentally different types of cloud – layered and convective, and only two basic constituents – water droplets and ice crystals. The formation of cloud together with stability and instability are discussed in Chapter 3.

Layered clouds are much greater in horizontal than vertical extent, usually with more or less uniform base and top; they can extend for hundreds of kilometres, especially where associated with weather fronts, and develop when the atmosphere is stable.

Convective clouds are generally discrete and cellular, their depth usually equal to and often much greater than the distance across. The base is quite uniform, except sometimes when cloud is dissolving, but tops vary considerably with turrets of cloud projecting above the main mass. They form only when the atmosphere is unstable.

The main cloud categories are shown in Table 6.1, although they may be subdivided further. Cloud nomenclature follows schemes derived by botanists and biologists to categorise species of plants and animals using Latin names, but it is a good deal simpler! The system for clouds was derived by Luke Howard, (a London pharmacist) in 1803.

Table 6.1 The ten main cloud categories and abbreviations

	Height of base	Convective	Layered
Low (C_L) (formed of water droplets)	0–2.5 km (0–8000 ft)	Cumulus (Cu) Cumulonimbus * (CB)	Stratus (St) Stratocumulus (Sc) Nimbostratus * (Ns)
Medium (C_M) (water droplets and ice crystals)	2.5–6 km (8000–20 000 ft)	Altocumulus (Ac) **	Altostratus (As) Altocumulus ** (Ac)
High (C_H) (ice crystals)	6 km (20 000 ft) or more	Cirrocumulus (Cc)	Cirrus (Ci) Cirrostratus (Cs)

Notes: * Deep clouds with ice crystals at higher levels.
 ** May be convective or layered.

Low clouds are arbitrarily specified by international convention as being those based below 2.5 km (8000 ft); high clouds have their bases from 6 km (20 000 ft) upwards, and medium clouds are those between the two.

Table 6.2 Common descriptive terms for clouds

Altus (alto)	Height	Used to distinguish cloud in the medi um height band (see Table 6.1)
Capillatus	Having hair	Applied to cumulonimbus where the top of the cloud has the typical fibrous character of ice cloud, distinguishing it from large cumulus
Castellanus	Castle-like battlements	Applied to altocumulus when it has the appearance of high-level cumlulus cloud
Cirrus (cirro)	Lock or tuft of hair	
Congestus	Piled up	
Cumuliform	Heap (heaped)	
Floccus	Tufted	Usually applied to cirrus which has a tuft or turret, with wisps trailing below
Fractus	Broken, irregular or ragged	Usually applied to stratus which forms below rain-bearing cloud
Humilis	Flattened	Used to describe shallow fair-weather cumulus
Lenticularis	Lens or almond-shaped	Describes many wave clouds
Mamma	Udder or protuberance	Form when cloudy down draughts push into dry air below the main cloud base and evaporate from the edges, forming a rounded, inverted dome-like shape. Usually associated with cumulonimbus
Nimbus	Rain cloud	
Pileus	Cap or hood	Forms where a large convective cloud deflects a moist layer above sufficiently to produce a separate cloud, rather like an inverted saucer. Sometimes two or more pileus clouds form.
Stratiform	Flat (flattened)	

It is not always possible to decide a cloud type, because there is inevitable overlap. This is due to the arbitrary subdivision of cloud base height, and also to the occasional hybrid cloud which is partly convective and partly layered. Probably the most difficult area is where one cloud evolves into another, especially the progression from cumulus to cumulonimbus where, viewed from below, it is often impossible to distinguish between them.

The ten main cloud categories

The ten main cloud categories listed in Table 6.1 are described here, together with wave clouds; further descriptions are often applied in particular circumstances, the most common of which are listed in Table 6.2.

Stratus (St)

This is a uniform, featureless cloud usually with a base below 500 m (1650 ft), always formed entirely of water droplets. It is sometimes shallow and tenuous, especially where it results from fog lifting as a morning breeze picks up. Often moist winds blowing over coasts or hills are lifted sufficiently to form stratus and hill fog (which amount to the same thing). Warm winds over a cold surface, especially lying snow, produce extensive stratus, similar in nature to advection fog but lifted above the surface. On radiation nights when the low-level air is moist and there is a breeze, low stratus tends to form rather than fog. This is because eddies or turbulence from the wind mix night-time cooling at the surface through a depth of a few hundred metres, and as the air cools this is eventually above its condensation level, and stratus forms. When rain or snow falls through clear air beneath its parent cloud, it sometimes cools moist layers below their dewpoint producing ragged patches of stratus below the main cloud base; this is called **fractostratus**.

Stratocumulus (Sc)

Perhaps the most common cloud, stratocumulus frequently forms in the lowest few kilometres when wind-driven turbulence lifts air above its condensation level. Stratocumulus displays a characteristic banded structure, these bands usually lying across the wind direction. Time-lapse photography shows that these bands tend to dissolve on their downwind side and form upwind; hence they appear to roll slowly across the sky. Sometimes blue sky can be seen between, but often only the gradation of dark and light grey strips shows the variation in depth. Stratocumulus frequently occurs in several layers, or in combination with altocumulus. It is formed completely of water droplets and, although not itself associated with rain, may markedly increase rain falling through it from higher cloud – this is called a **seeder/feeder** process. Sometimes stratocumulus forms from the spreading out of cumulus clouds. Frequently it is found just below a temperature inversion, where either a layer of moist air cools by radiation (usually at night), or convection currents lift moist air from below, forming cloud which is forced to spread out under the inversion (usually by day). In all these cases, turbulence in the windflow gives the cloud layer a typical roll structure.

Cumulus (Cu)

Cumulus is the most common convective cloud and forms only when the atmosphere is unstable (see Chapter 3). A warmed bubble or plume of air forms cloud as it rises above its condensation level, and continues upwards until it runs out of buoyancy. Cumulus clouds tend to be discrete, because convective plumes break away from the surface individually. Also there are surrounding clear downdraughts to replace the ascending air.

Small cumulus clouds form when the unstable layer is shallow, with plenty of blue sky between. They are called fair weather cumulus, or cumulus humilis because, not only are they unable to produce showers, but they indicate that deeper shower clouds are unlikely.

Large cumulus or cumulus congestus develop when the unstable layer is deeper than about 3000 m (10 000 ft), and where they rise above the freezing level they give showers. They have large, rounded, well-defined turrets, some of which may be decaying while others grow. From beneath they appear dark grey or almost black, and often little or no blue sky can be seen between. Tops are sharply outlined.

Cumulonimbus (CB)

These are the deepest and most vigorous convective clouds, and produce not only showers but thunderstorms, hail, squally winds and occasionally tornadoes. Sometimes the strength of the deep convective upcurrents push cumulonimbus tops well into the lower stratosphere.

Cumulonimbus like cumulus, usually form by heating from below, but the most intense cumulonimbus are helped by cold winds at high levels making the atmosphere even more unstable. Tops are predominantly of ice crystals with a hazy outline, distinguishing them from large cumulus. Often the icy top of a mature cloud takes on a characteristic '**anvil**' form as it spreads out at the tropopause. Lower down they are a turbulent mixture of ice and water, containing rain, hailstones and snow.

Cumulonimbus are avoided where possible by aircraft because strong up and down currents within and near them create severe turbulence, while their high water content can rapidly produce thick layers of ice on cold airframes.

Altostratus (As)

This layer cloud is based at medium levels. It forms by the slow ascent of air over a wide area, especially ahead of a warm front or occlusion where it is often a precursor of rain. Often it first appears as a thin, misty cloud, but as it thickens becomes a dark grey, uniform sheet. Frequently, it coexists with layers or patches of altocumulus or statocumulus beneath, and merges with cirrus or cirrostratus above. Although composed mostly of water droplets, thick altostratus can produce light rain, though most of this evaporates before reaching the ground.

Nimbostratus (Ns)

This is deep and extensive layer cloud from which rain (or snow) is falling. It often represents an evolutionary step following thickening altostratus near a weather front. The first rain from altostratus evaporates into clear air below, but soon the cloud base lowers and nimbostratus forms with rain reaching the ground. Nimbostratus appears black from below, but as rain or snow becomes heavier the base may become indistinguishable, with just ragged scuds of low fractostratus to be seen. This cloud has a high water content because of continued ascent, and can give severe airframe icing.

Altocumulus (Ac)

In most of its several forms, altocumulus is simply a medium-level version of stratocumulus. It forms in rolls or patches, with or without gaps between, by turbulent mixing often in moist layers remaining from dispersed fronts. Sometimes, several layers can be seen at once. Like stratocumulus it can form from the spreading out of cumulus tops or decaying cumulonimbus, especially late in the day when daytime convection ceases. More significantly to a forecaster, it sometimes takes the form of cumulus with a high base, in which case it is called altocumulus castellanus. However, unlike cumulus which forms by convection from heating at ground level, altocumulus castellanus usually forms when cold middle-level winds overrun warmer low-level air to produce an unstable layer at medium levels. This is frequently the first visible sign of the imminent thundery breakdown of a summer hot spell. Occasionally rain (or a shower) falls from altocumulus, but it is invariably light and often evaporates long before it reaches the ground. Sometimes this can be seen as streaks or **virga**, streaming from under the cloud.

Cirrus (Ci)

This is high cloud composed entirely of ice crystals and takes many forms. It is non-uniform and often thin and wispy, sometimes with thicker bright twisted sheaves, and all shapes between. It forms by ascent in the upper troposphere, often with no associated low-level front or weather system involved. Sometimes it can be traced back to long-decayed fronts or the icy tops of cumulonimbus clouds. Occasionally it is 'manufactured' when condensation trails from high-flying aircraft 'seed' already moist or very cold air. In general, cirrus, being high and tenuous, does not greatly reduce incoming solar radiation; neither does it materially inhibit overnight surface cooling.

Condensation trails or **contrails** form to the rear of high-flying aircraft, either from condensation into air saturated as it cools due to the pressure reduction to the rear of wings and tailplane, or from water vapour in the hot exhaust gases condensing as it mixes with cold ambient air. When the air is dry contrails soon evaporate, but if it is moist or extremely cold they persist for hours or even days. Sometimes they seed further cloud and become several kilometres wide, appearing on satellite pictures like a criss-crossing railway junction.

Contrail forecasting was important in wartime when highflying aircraft wished, as far as possible, to avoid detection.

Cirrostratus (Cs)

Unlike cirrus this high, diffuse ice cloud takes the form of a uniform sheet through which the sun or moon may be detected, sometimes with halo effects. It is produced by the slow ascent of air and condensation or sublimation high in the troposphere, usually well ahead of weather fronts. It appears in meteorological folklore because it is often an early indication of rain (see Chapter 9).

Cirrocumulus (Cc)

This attractive ice cloud is the high-level equivalent of stratocumulus or altocumulus, and much less common than either. It is formed of cells aligned in streets rather like ripples in the sand on the beach; usually blue sky can be seen between. The streets themselves may be inclined at angles to one another, making interesting patterns. A **mackerel sky** is a particularly beautiful example, reminiscent of the delicate bone structure of a fish, with 'V's inset one inside another across the sky. Cirrocumulus forms by wave motion or turbulence through a moist layer in the high atmosphere; winds with marked differences in direction at the top and bottom of the layer produce quasi-uniform patterns of up and down currents, which become visible only when clouds form.

Wave cloud

This is not in the main cloud classification of Table 6.1, and yet is important for many reasons. When windflow is deflected upwards by a range of hills, then provided the air is stable it will tend to return to its original level, although a patch of cloud may form over the mountain tops. If the air is unstable, cumulus or even cumulonimbus may form; this is one reason why showers are more common in hilly areas. However, if there is a stable layer with less stable air above and below, this is displaced upwards but instead of just descending to its original level it overshoots; it then returns upwards overshooting again and so on (see Figure 6.3) The result is an oscillatory wave motion in the airstream downwind of the hills. Clouds form in the wave crests if the air ascends above the condensation level, and they stay 'fixed' relative to the hill or mountain, hence are called **standing waves**. We thus have a paradox of a cloud, or a whole 'train' of wave clouds, apparently motionless in the sky but with strong winds blowing through them. Waves are much appreciated by glider pilots because they provide sustained uplift.

Wave clouds are normally classified as stratocumulus or altocumulus, depending on cloud base height. Wave motion often extends high above the hill or mountain tops; indeed it is sometimes seen in cirrus clouds, especially downstream of major mountain barriers such as the Rockies, Andes, Alps and Pyrenees.

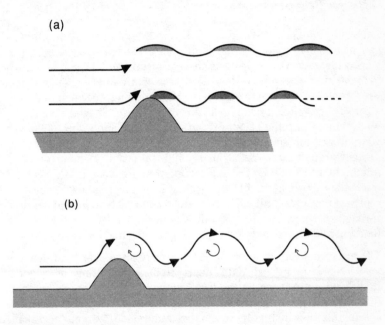

Figure 6.3 Wave motion and clouds.

(a) Often winds blowing across a range of hills produce wave-like motions to the lee, with cloud forming in the wave crests. These may be at two or more levels and remain stationary.

(b) Even if the air is too dry for cloud to form, turbulent air currents may be generated which can endanger small aircraft.

—— Direct upper air observations ——

The atmosphere is emphatically three-dimensional, and any attempt to forecast the evolution of weather requires observation of winds and temperatures in the upper air, where the main transport of energy around the globe occurs. Recent advances in forecasting lean heavily on developments in computing and mathematics, but there is no doubt that more accurate upper air observations, especially from aircraft, have played an important part.

The radiosonde

Following the Second World War, with the rapid increase of commercial aviation, a great deal of effort was expended in establishing a worldwide network of upper air observing stations. These release balloon-mounted packages of instruments or **radiosondes** which measure temperature, humidity and wind through the troposphere into the stratosphere, often reaching 30 km (20 miles). They transmit the observations by radio back to the station. Today and every day of the year, upper air soundings by radiosonde are carried out simultaneously around the world at nearly 500 stations at 12-hour intervals (midnight and 12:00 UTC). Information from each sonde is coded and immediately transmitted to the nearest meteorological data collecting centre. From these, all such data are sent to meteorological services throughout the world. Radiosonde readings have been and remain a fundamental component of the global observing and forecasting effort.

The great majority of upper air stations are sited over land, leaving vast oceans empty of data apart from a speckling of islands. Radiosondes are flown by the few remaining weather ships, but that fills only one or two gaps. An exciting recent development has been the installation of highly automated radiosonde systems on ocean-going cargo vessels. These require no intervention by the crew of the vessel, even the data obtained are disseminated directly via satellite.

Radiosondes measure pressure, temperature and humidity using elements whose electrical properties vary, and which can, therefore, easily translate into radio signals.

- Pressure is measured by aneroid capsule of the type described earlier.
- Temperature is measured by **thermistor** which is a semiconductor sensitive to changes over a wide range.
- Humidity is measured by an hygristor as described earlier. Originally gold-beater's skin was used. **Gold beater's skin** is part of the membrane from the intestine of an ox, which was used by beaters of gold leaf. It expands with increasing humidity and is also strong.

- Wind speed and direction were originally determined by tracking the balloon by radar, which required a large metallic reflector to be carried. This meant additional weight and consequently more hydrogen and bigger balloons. In recent years, extremely precise position fixing has become possible using satellite systems, or surface-based radio navigation aids such as Loran. **Loran** is shorthand for Long Range Navigation; it is one of several systems that rely on the precise measurement of synchronised radio signals from several widely separated stations to accurately compute geographical position. This enables the average wind to be calculated over short periods of time simply from the accurate measurement of successive positions of the instrument.

When the balloon carrying a radiosonde eventually bursts, the instrument package gently returns to Earth by parachute. Occasionally, one hears of mysterious objects being discovered on roofs, or frightening cattle and these are often found to be radiosondes. They normally have a prepaid address label on them and in some countries the finder will recieve a small reward. In general, though, they are used once only.

Radiosondes are launched from the surface, but **dropsondes** travel in the opposite direction. They descend by parachute from research aircraft or more rarely a rocket, but are expensive to use and hence are primarily research instruments. For example, investigations into the detailed structure of weather fronts have involved the launch of many dropsondes over a small area nearly simultaneously, to obtain fine detail of temperature, wind and humidity structure.

While the radiosonde is the mainstay of accurate upper air measurement around the globe, it has limitations. It does not ascend vertically, and readings are spread over about an hour. Small errors and uncertainties due to these factors have not mattered in the past, because they have been negligible in comparison with large gaps in our knowledge. However, as forecasting improves, the quality of data assumes increasing importance, and soon radiosonde observations will need to include adjustments for spatial displacement from the launch station, and time difference from the nominal time of ascent. Occasionally a balloon rises through a local convective cloud, which means its observation is interesting but quite unrepresentative of the surrounding atmosphere. The vigilant forecaster will intervene to ensure such data are not misinterpreted, either by colleagues or computer!

Aircraft reports (or Aireps)

Direct in-flight measurements of temperature, pressure and wind by commercial aircraft has increased considerably in accuracy and quantity in recent years, contributing greatly to improving forecasts from which all, including the airlines themselves, benefit.

- Temperature is measured by thermistor incorporating a correction for frictional heating due to the speed of the aircraft.
- Humidity is not measured routinely.
- Wind speed and direction are calculated by subtracting the aircraft's groundspeed from its airspeed again using accurate position fixes from external satellite or Loran navigation systems.

Most long-haul aircraft transmit meteorological reports every 5 degrees of latitude or longitude at cruising level, but many are having automatic systems installed. These take readings every few minutes from take-off to landing, which are automatically transmitted via satellite to meteorological collecting centres. They are known as **ASDAR** systems, abbreviated from 'aircraft to satellite data relay'.

Instruments mounted on kites, rockets, gliders and constant-level balloons circling the globe have all had a part to play in the past acquisition of upper air data. Now they have almost completely given way to cheaper more reliable observations from radiosondes or commercial aircraft, together with increasing amounts of valuable data obtained by remote sensing.

———— Remote sensing ————

Until recently, measurements were taken at a point, but regarded as representative over a much wider area. For example, a thermometer measures the temperature of only a small volume of air, but is often regarded as a meaningful guide for a town or even a country. In the past few decades there has been an explosion in information obtained remotely, sometimes by instruments thousands of miles away, and often this relates to an average value over a large slice of atmosphere.

All **remote sensing** systems rely on the measurement of a reflected and often minute signal usually returned from a transmitted beam. The beam might be comprised of microwaves, sound waves or even laser light; alternatively the signal may be from reflected parts of the sun's radiation or radiated by gases in the atmosphere. As techniques have become more refined their value has increased, both from ground-based instruments and satellites, because they enable measurements over a large area that would otherwise require a huge and expensive network of reporting stations. Furthermore, much of the information is available to the user either immediately or soon after the time of observation.

Imagine bouncing a ball from a wall. Its return path can be anticipated by eye from the speed with which it is thrown, its angle of impact, the distance from the wall, and the characteristics of the ball itself. What would happen if the wall was invisible? The ball would act in exactly the same way, and we could deduce a great deal about the position and character of the wall by observing the motion. Remote sensing is like that; we are sending the equivalent of millions of minute balls into the atmosphere in known directions and accurately timing and counting those (if any) that return.

Weather satellites

Artificial satellites are launched into orbit around the Earth in such a manner that the centrifugal force outwards is exactly balanced by the world's gravitational pull. Since they are above the atmosphere there is no air resistance or friction to slow them down. Their height is determined by the velocity with which they are launched into orbit. The direction of launch, and the time of day the satellite enters its orbit, dictate which parts of the Earth will be traversed and when.

Various types of satellite instrumentation are used to provide meteorological data, and remote sensing by satellite is rightly acclaimed as one of the greatest advances ever in scientific observation. On the other hand, many of the data obtained are coarse in resolution both in space and time, and must be interpreted with care. The latest numerical models of the atmosphere (see Chapter 7) are not helped by poor-quality data, and it is essential to continually review and refine what information is fed through and used in compiling detailed meteorological analyses. Certainly the most striking and immediately useful data for a human forecaster are satellite pictures, which can be obtained by anybody with the appropriate receiver, or over the Internet.

There are two main types of meteorological observing satellite, each of which has advantages and disadvantages in the cloud imagery it can provide.

Polar-orbiting satellites

As the name suggests, these take a path almost directly over both poles. The orbit is, of course, fixed in space but, since the Earth is rotating on its axis, the satellite 'sees' a different strip underneath it on each pass (see Figure 6.4). Orbits are usually chosen in such a way that the satellites 'follow the sun'. The higher the satellite, the longer each orbit takes and the wider its field of view (and, other things being equal, the longer it will stay aloft). However, lower satellites resolve more detail in pictures. American polar-orbiters are usually at a height of about 850 km (530 miles), making one orbit every 1 hour 42 minutes approximately, with each pass about 25 degrees of longitude west of its predecessor. Russian polar orbiting satellites have tended to be about 400 km (250 miles) high. Composite pictures are often compiled from successive orbits of a single satellite, but cloud is disjointed across the orbit boundaries where movement has occurred between times.

Geostationary satellites

Launched in the same direction as the Earth rotates, these take up a position above the Earth's equator. Furthermore, the height is selected such that each orbit takes 24 hours. This means each satellite maintains a fixed position relative to Earth, directly above a point on the equator. The height of a geostationary satellite is necessarily 36 000 km (22 500 miles); if lower it will orbit too fast and vice versa. The fixed position of a geostationary satellite enables time-lapse sequence of pictures to be made, showing the past movement and evolution of cloud. However, the high altitude means that less detail can be seen than from polar orbiters. Also the curvature of the Earth means that little is clearly visible polewards of about 65 degrees of latitude. Five strategically placed geostationary satellites maintain a constant vigil over nine-tenths of the globe, including the vast data-sparse oceans of both hemispheres (see Figure 6.4). This is of inestimable value because it includes the breeding grounds of hurricanes, enabling them to be detected earlier than was previously possible.

Satellite instrumentation

Although satellite pictures are the most commonly available product, no meteorological satellite carries a camera. All information is obtained

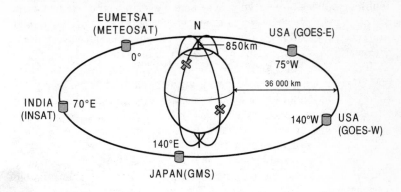

Figure 6.4 Five geostationary satellites each view fixed segments of the Earth from their fixed positions (relative to the Earth) above the equator at a height of 36 000 km; together they span the globe. Polar orbiting satellites are much nearer the Earth and view a varying narrow sector of the Earth's surface as it rotates beneath.

by measuring radiation from each small sector of the Earth and atmosphere below in selected wavebands, by means of instruments called **radiometers**. The most familiar waveband to us is visible light, which simply comprises reflected sunlight. Pictures composed using this look just like photographs and, unsurprisingly, are known as **visible imagery**. Of course, this does not mean that the remainder are invisible, at least not after processing!

There are problems with visible imagery. Occasionally, the sun reflects from a nearly calm water surface giving a bright white image, called **sun glint**, which can be confused with cloud, though normally the shape of the reflection or an intersecting coastline gives the game away. Snow is also a brilliant reflector of light, and at high latitudes can be as bright as high cloud. Here again usually shadows, coastlines or warmer river valleys distinguish it from cloud; snow over mountains often has a characteristic filigree pattern. More important, shallow low cloud is just as bright as deeper and higher rain-bearing cloud. It is not always easy to discriminate. The most serious disadvantage, however, is that visible imagery is impossible at night, which means half of the world cannot be seen at any one time.

These problems are largely overcome by using other radiometers tuned to measure radiation in the longer wave infrared (IR) portion of the spectrum. **Infrared imagery** is effectively a measure of temperature variations beneath the satellite, which of course exist at night as well as by day. High cloud tops are very cold, and translate as bright white on IR imagery, while lower, warmer clouds appear as progressively darker shades of grey. The Earth's surface ranges from black on a hot day to light grey on a cold night. But beware, pictures are often coloured or shaded differently using computers, to emphasise differences or to make them more attractive; bright, cold, high cloud becomes dark threatening cloud on a TV weather forecast! IR imagery has two main disadvantages: first, pictures are less sharp than visible images; second, low (warm) cloud tops are often impossible to distinguish from the surrounding land or sea. This second problem can be solved when a combination of visible and IR imagery is available, because all cloud will be lighter than sea or land on visible imagery.

Other wavelengths are sampled by satellites. Water vapour in the upper atmosphere radiates strongly in certain wavebands, and using these it is possible to 'see' areas of moist air even where cloud is absent or, equally important, deep dry air. Often whirls of high-level moisture pick out the structure of a major weather system better than more chaotic visible cloud.

Other gases, notably carbon dioxide (CO_2), radiate strongly at certain wavelengths, and those wavelengths vary with pressure, which itself decreases with height through the atmosphere. Furthermore, the intensity of radiation depends on temperature. This sounds like a meteorologist's dream, because it gives a means of remotely measuring temperature over a wide range of heights. If we could 'chop' the radiation spectrum into narrow wavebands, relate each to a height, and measure the intensity of radiation from CO_2 in each, then it should be possible to derive a temperature profile. The earliest instrument for doing just this was in fact called a '**chopper radiometer**', and upper air temperature measurements by satellite, or **satems**, have been made for many years. Unfortunately, unlike measurements by radiosonde or aircraft, these are average temperatures, each relating to a box of atmosphere several kilometres deep and tens of kilometres across. While they are of value over data-sparse oceans, they are a poor substitute for accurate point measurements.

Displacement winds (satobs)

Using a sequence of satellite pictures it is possible to time the progress of a patch of cloud over a period, and hence deduce the upper wind. This has been done for several years but the results are not accurate. Cloud rarely, if ever, travels exactly with the wind. Even gradual ascent or descent significantly extends or evaporates the leading (or trailing) edge so that progress between two separate pictures is deceptive. Indeed, sometimes cloud is seen to disperse over a huge area over a short period of time, whereas at other times narrow streaks of high cloud (or jetstreaks) form rapidly along a path of several hundred miles. Few would misinterpret either of these as completely wind-driven, but unfortunately beween these extremes it is all too easy to be misled. In addition, there is always uncertainty as to the exact level of the cloud being observed. The height of shallow cumulus tops in the subtropical trade winds can be deduced fairly accurately, but at high levels satobs are often less accurate than short-period forecast winds by computer.

Scatterometer winds

In the mid 1990s an exciting innovation enabled surface winds over the sea to be observed from polar orbiting satellites, a truly remarkable achievement. The Earth Resources Satellite (ERS1) launched in 1994 carried the first operational **scatterometer**. This directs a microwave signal downwards to span a 500 km-wide (300-mile wide) strip of the Earth's surface. Each portion is viewed from three directions, by one of three aerials: first ahead of the satellite, then directly beneath, and finally to the rear. Differences in the reflected or back-scattered signal depend on the roughness of the sea surface (not large waves or swell). This roughness is closely related to the surface wind and direction, which are deduced and transmitted to meteorological centres. The instrument is similar to radar discussed below.

This system has been found to give excellent measurements of wind speed up to about gale force, above which the sea is so disturbed the signal becomes confused. The observed wind direction does not suffer this limitation, but has the drawback of being ambiguous; the line along which the wind is blowing is clearly defined, but not which way along the line. This fundamental uncertainty is not as serious as it might seem. Wind direction is almost always known approximately, and the correct alternative is normally easily deduced.

A scatterometer can 'see' through cloud which means it is of great value in locating centres of depressions, and especially in pinpointing hurricanes (see Figure 6.5). One of the ironies of improved forecasts has been that shipping is now directed away from areas of severe weather, denuding weather charts of observations near hurricanes. Apart from this, swathes of reliable wind observations over the vast oceans, largely devoid of other information, are a major advance in the quest for a complete knowledge of the atmosphere, so necessary for numerical forecasting, discussed in Chapter 7.

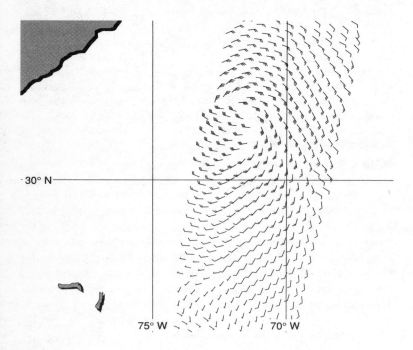

Crown copyright

Figure 6.5 Scatterometer winds. The flag-like symbols are plots of surface wind speed and direction observed by satellite using the disturbance of the sea surface. Each flag points in the direction from which the wind is blowing; a full feather represents 10 kn and a half feather 5 kn.
This orbit fixed the position of hurricane Emily precisely, seeing through cloud. Indicated wind speeds near the centre are, however, much too light.

Weather radar

Since the 1940s radar has evolved from a crude aircraft detection and tracking system, into a multitude of sophisticated instruments, tuned to detect selected 'targets'. In meteorology those 'targets' are now almost invariably rain, snow or hail, with a single well-sited radar able to measure their location and intensity within 200 to 300 km (125 to 200 miles) of the observing site. **Rainfall radar** imagery has become so valuable that many countries have networks of radars enabling a composite near instantaneous rainfall 'picture' to be derived, covering many thousands of square kilometres. Time sequences of these composite images give an often dramatic 'movie' of the progress and evolution of areas of rain. They have become an essential tool for short-period forecasting, and are used widely in TV weather presentations. It may even be possible soon to be able to see up-to-date rainfall imagery on your wristwatch!

The usual meteorological radar system comprises a transmitter sending a succession of pulses of high-frequency radio waves, and a sensitive receiving aerial to detect the lapsed time and intensity of the returned echo. The position and nature of all returns are automatically computed, and usually displayed on a cathode ray tube or monitor. Normally, a rotating aerial provides continuous details of precipitation falling over a circular area. This is a plan position indicator or PPI system. Less common is a nodding aerial sampling a vertical slice of atmosphere in one direction, the range height indicator or RHI system.

A radar beam may be reflected by hills or mountains producing permanent echoes, and partly shielding rain beyond from view. Thus, the ideal site for rainfall radar is on top of a hill or ridge, unobstructed by other hills or nearby buildings.

The extent by which microwaves are reflected by water droplets depends on the wavelength of the radar beam, and increases rapidly with drop size. A drop twice the diameter of another contains eight times as much water, but reflects no less than sixty-four times the microwave energy. Wavelengths of between 3.5 cm and 10 cm (1.5 and 4 inches) are normally used for rainfall radar, because within this waveband cloud droplets are effectively transparent and drizzle will only produce a signal if it fills most of the radar beam. Even small showers containing large raindrops may be detected over 100 km (60 miles) away, and large storms or active fronts further away still.

A radar 'picture' gives an instantaneous snapshot of falling rain, and is extremely valuable for fixing the position of fronts and individual storms. Time-lapse sequences give valuable insight into the development of fronts, frontal waves and storms. In the United States studies have found that the radar 'signature' of heavy rain associated with a severe storm differs from that for ordinary thunderstorms, having a characteristic comma shape and a clear 'dry slot'. This echo often provides the first indication of the development of tornadoes, enabling them to be tracked and timely warnings issued (see Chapter 10).

It is difficult to measure rainfall amounts accurately by radar. Every raindrop in the beam prevents a small part of the beam from penetrating further – just like torchlight being reflected and scattered as it is shone into fog or smoke. This means that a large area of rain, especially heavy rain, will register only partly on the radar screen. It is possible to correct for this attenuation electronically, but not always accurately. Furthermore, not all rain detected by radar reaches the ground; some evaporates, especially if the air under the cloud is dry. Often well ahead of a warm front, rain initially starts to fall from cloud above 3 km (2 miles) and totally evaporates before reaching the ground. Nevertheless, despite difficulties, rainfall measurement by radar is a valuable tool. Where it is augmented by even only a few conventional rain gauges, the radar image can be calibrated, and the whole area of echo interpreted more accurately. Rainfall radar calibrated in this way usefully measures the extent and variability of rain on scales that would require vast numbers of conventional gauges. Consequently it is widely used by river and water authorities to predict river flow and run-off, sometimes enabling flooding lower down a river valley to be predicted, and preventive measures to be taken. There is untapped potential for rainfall radar sequences to be made available in real time to the general outdoor public, perhaps through the Internet.

Other sources of serious error in rainfall radar measurement mainly relate to the non-uniform transmission of microwaves through the atmosphere. A radar beam can be deflected by sudden changes of temperature and especially humidity along its path – just as light is bent as it passes through a lens. Often in anticyclones there is a sharp increase of temperature through a layer above the surface (subsidence inversion) and an abrupt decrease in humidity. This can cause a radar beam to be deflected downwards into the ground, producing a bright return when there is neither rain nor the prospect of any.

Similar conditions lead to poor television reception for the same reason. Less often a radar beam becomes trapped within a narrow layer of the atmosphere and reflects from mountains or rain-bearing clouds far beyond its normal range. These strong but late return signals may be misinterpreted as short-range echoes from later pulses. Entrapment of a beam in this way is called **radar ducting** and the layer of the atmosphere a **radar duct**. Frequently over warm seas there is a sharp boundary between low-level moist air and much drier air above. This can prevent a radar beam pushing through and 'seeing' distant storms; it is called an **evaporation duct**. All spurious echoes, and fortunately most of them are obvious to an experienced eye, are referred to as **anomalous propagation** or **ANAPROP** for short.

The crystalline structure of ice, even in the form of large snowflakes, does not reflect radar nearly as well as raindrops, and in general snow gives a weak radar return. However, where snowflakes start to melt as they fall into warmer air, the outer coating of water makes them appear to radar as extremely large raindrops. The result is a very bright echo, often taking the form of a line a few kilometers wide within an area of rain. This is known as a **bright band** and the process as **bright banding**. It can easily be misinterpreted by an inexperienced operator as a narrow zone of heavy rain.

Wind profiler

The problem of radar ducting, described above, has been elegantly turned to advantage by using small-scale variations of temperature and humidity in the vertical to deduce wind speed and direction. The wind profiler comprises a longwave microwave beam directed vertically, with two further beams each slightly out of the vertical and at right angles to one another, so the layout is like an 'L'. Any small-scale disturbance detected by the main beam can be tracked by the others as it is blown along by the wind, and hence the height of the disturbance, together with its speed and direction of movement, can be computed, including its vertical velocity (ascent or descent). Profilers can be augmented by radiometers to measure temperature, in the same way as from satellites, as described earlier. Accuracy is substantially inferior at present to that of the radiosonde, and the equipment is expensive; on the other hand it enables a continuous record to be obtained.

Acoustic sounder

This instrument is a poor relation of the profiler discussed above. It measures the reflected signal from vertically directed sound waves, which varies due to change of temperature and humidity with height. In particular, reflection is enhanced by temperature inversions. The 'echo' pattern at the ground is related to the level reached simply by timing the delay between sound and its return. Variations of echo intensity provide indications of the height of the top of fog and the thickness of low cloud, helping a local forecaster to predict the time of clearance. Even details of temperature inversions can be detected and studied.

The siting of acustic sounders is difficult because of the effect of extraneous sound. Measurements are largely qualitative and difficult to use operationally. Their role is primarily in research studies, especially into the formation and decay of fog and low cloud, and into pollution potential as low-level inversions develop and decay.

7

FORECASTING THE WEATHER

In Chapter 5 we considered the main weather systems of middle latitudes: lows or depressions with their weather fronts, which bring unsettled and sometimes stormy weather; and highs or anticyclones when it is usually fine and dry with light winds. We are now going to look at the major problem of meteorology which is how to forecast them and the weather they bring. In doing this it is important to remember that, just like human beings, no two weather systems are ever precisely the same.

Weather forecasting first requires understanding of how the atmosphere works, second, it needs knowledge about current weather patterns, and third, the ability to put both of these together to determine future developments.

Until the 1960s forecasts were produced manually. Observations collected from, at most, one hemisphere were plotted on charts. They were carefully analysed by drawing isobars, and using these and weather reports to position weather systems and fronts. Forecasters would also have charts plotted for higher levels using observations from radiosondes and aircraft, so that upper air temperatures and winds, especially the location of jetstreams, could be determined. A combination of techniques would then be used to forecast the equivalent charts in the future, usually for 12, 24, 36 and 48 hours ahead. Most of the forecasting methods were subjective, and experience was valuable. First, current trends in the motion and development of systems were assessed, and future progress modified in the light of the changing upper winds across them. Then various rules would be applied, some based on research and others on statistics, to produce a set of forecast charts. Finally, the weather over the forecast period likely to be associated with these charts was assessed and forecasts compiled accordingly.

Procedures were mostly graphical and qualitative. It was particularly difficult to predict when and where the next depression was likely to form, and precisely in what direction it would move. This is critically important because if forecasts of large-scale weather systems and their associated fronts are poor, then the timing and intensity of small-scale detail such as areas of cloud, rain and showers will often be completely wrong.

At that time weather forecasts up to about 24 hours ahead were often useful, but beyond that they were completely unreliable. A present-day forecast for three days ahead is (on average) more accurate than a one-day forecast was in the 1960s, a tremendous advance that has brought great benefits to society. What lies behind this revolution? The answer is numerical weather prediction or NWP for short.

Numerical weather prediction and numerical models

Numerical weather prediction is forecasting weather by simulating the atmosphere using systems of mathematical equations known as **numerical models**. It has almost entirely supplanted earlier subjective methods, and is now universally acknowledged to be the only worthwhile method for forecasting usefully beyond a day or so. Success has been achieved by intense meteorological research, but it would have been quite impossible without the concurrent development of increasingly powerful computers, far beyond the dreams of at least one young meteorologist in the 1950s!

The first attempt at NWP was carried out by L F Richardson in 1922. Although unsuccessful due to the paucity of initial information and the lack of computing power, he led the way in showing that it was possible.

What is a numerical model?

Any simple system such as a bouncing ball or vibrating violin string can be expressed mathematically. If we wish to know how high a ball will bounce when dropped on a hard surface a mathematical equation can tell us, provided we supply a few details like how bouncy the ball is, and what height it is being dropped from. The musical note a violin plays when a bow is drawn across it can similarly be determined from a simple equation involving the length and character of the string. Both these equations are simple **mathematical models**, because they simulate the way the ball or the violin behaves. By solving them with the appropriate initial conditions, they can be applied to any bouncing ball, and any stringed instrument.

The laws of physics and mathematical equations governing the motion of fluids have been well known for over a century, but the complexity of our earth/atmosphere system is daunting. It is a thin skin of air on a rotating sphere, unequally heated by the sun, part sea, part land, with mountains, deserts and forest. We know the gas laws, the laws of motion and gravity, conservation of energy and matter; we understand processes involving water in the atmosphere, incoming and outgoing radiation and the geometry of the Earth's orbit and rotation. The equations encompassing all these laws and processes together form a **numerical model of the atmosphere**.

Individually most of the constituent equations are reasonably straightforward; taken together they are exceedingly complex. Furthermore, there are many uncertainties and approximations which meteorologists continuously seek to resolve. Clouds form and dissolve governing how much of the sun's heat penetrates to the ground. Where snow lies on the ground it reflects much of the sun's heat; water evaporates, rain falls, fogs form and dissipate.

Even if we knew absolutely everything about the atmosphere at some point in time, our large set of equations is quite impossible to solve directly to see how things evolve. Other techniques have to be used, with simplified equations which are solved over short time intervals, at a network of discrete points across the globe.

Our model atmosphere inside the computer is split up into 'parcels' of air, stacked in columns, with thousands of these columns distributed over a grid of points (**grid points**) spaced over the Earth's surface.

First, we determine from all available information what the wind, temperature and humidity are within all of these parcels at a fixed time; this is our **model initialisation**. The model equations are then applied to each 'parcel' in turn, using what is happening at surrounding grid points to determine changes over a short fixed time interval called a **time step**. Does a 'parcel' expand or contract? Is the air moving through it rising, descending, changing direction or speed? Does it become saturated or are water droplets within it tending to evaporate? Is it clear or cloudy? Does it become warmer from compression or incoming heat from above or below? Does it cool from expansion, outgoing radiation, evaporation of water or melting of ice crystals? The answers to these questions and many more influence what is happening at adjacent parcels in the column and also at nearby grid points, whose values are adjusted accordingly; but provided time steps are short those further afield will not be affected immediately.

The distance separating grid points is referred to as the **model resolution**, because, in general, a model cannot successfully carry within its simulated atmosphere any weather system smaller than twice this size. Unfortunately, to double the resolution (or halve the grid spacing) requires approximately eight times the computing power to carry out a forecast in the same time. This explains the eagerness of meteorologists to exploit the latest most powerful computers. The best NWP models now can carry hurricanes and predict their tracks and development, but certainly not individual storm clouds which give rise to tornadoes. The balance between model resolution and computing power is addressed at some meteorological centres by using a global model with a fairly coarse grid, but nesting smaller area, higher resolution models within it, to forecast for areas of most interest in more detail.

A fairly typical family of NWP models is run, in 1996, by the UK Meteorological Office. The Global version has 1 187 424 grid points with a resolution averaging about 100 km (60 miles). Within this the Limited Area Model covers only the North Atlantic and European sectors, but with a higher resolution of 50 km (20 miles) using 574 332 grid points. Both the Global and Limited Area Models have 19 levels in the vertical. The highest resolution Mesoscale Model is restricted to little more than the United Kingdom and surrounding sea areas. It has a resolution of 15 km (10 miles), 262 384 grid points and 31 levels in the vertical. Time steps used for the models vary from between 5 and 15 minutes, and the longest forecast period is 144 hours or 6 days. Simplified versions of the Global Model are used for climate modelling and research into climate change (see Chapter 11). These use much longer time steps and simplified equations enabling forecasts to progress through 100 years or more! Of course, by that time it is not the day-to-day weather that is of interest, rather the month by month, and year-by-year climate.

Computations proceed at a rate of several billion arithmetical operations per second, or gigaflops. The computation part of a global six-day forecast takes less than 20 minutes.

An arithmetic operation is called a **floating point operation** or **FLOP**; a million per second is a megaflop, and a billion per second is a gigaflop.

It may appear from this that the human forecaster has no part to play in present-day forecasting, but that is not so. Running a computer model is just one important step and, at least for the next few years, the winning team will remain a combination of human and machine.

The four stages of modern weather forecasting

The procedures leading to the production of a weather forecast may be conveniently subdivided into four main stages: initialisation, running an appropriate NWP model, outputting results, and interpreting and presenting them in a user-friendly form.

1 Initialisation

Few will not have heard the adage applied to computers – 'rubbish in means rubbish out'! Meteorologists are acutely aware that this can apply to NWP models and their forecasts. It is crucial that initial weather data are of high quality, so that knowledge of the atmosphere over the globe at the starting time is as accurate as possible.

The best first estimate (often called the 'first guess' field though this does it less than justice) of this initial state of the atmosphere is a short-period forecast from the previous run of the NWP model, called a **background field**. This is then modified in the light of actual weather reports.

Weather observations of all types are rapidly transmitted worldwide through cable and satellite links to a network of regional collection and distribution centres. In this way a comprehensive global weather data set is made available shortly after observation time to every meteorological authority. Surface and upper air observations, described in Chapter 6, are usually made at fixed times every three and 12 hours respectively. These are augmented by observations from many other sources such as aircraft and satellites that report continuously. Thus initialisation of global weather within the model occurs over a span of a few hours rather than at an absolutely fixed time. Most observations are transmitted in numerical coded form, and fed directly into computer systems.

Collecting data and inserting them into the model is not the end of the matter, errors must be corrected or eliminated. While a numerical model can itself detect many inconsistencies, and perform much of the initial quality control, ambiguity or conflict requires scrutiny by an experienced forecaster. Sometimes extra detail not directly available to the computer may also be incorporated. For example, the comparison of model background fields with satellite pictures, especially in data-sparse areas, may bring deficiencies to light. When this occurs fictitious observations may be devised and inserted to correct or improve initial fields.

Once initial pressure, temperature and wind fields have been deduced worldwide, values are allocated at every grid point and at all levels in the model for the initialisation or starting time, called T=0. Each value takes into account the background and all nearby observations, resolving inconsistencies and ensuring that the result is meteorologically 'balanced'. Those observations not made exactly at the initialisation time will be adjusted with regard to this time difference.

2 Running the NWP model

The model run itself is essentially a 'number crunching' exercise of breathtaking dimensions. A Global NWP Model run out to T+144 hours or 6 days uses several million individual observed items of weather information to initialise values at over a million grid points. The forecast run then involves about a thousand time steps, each requiring more than a billion individual arithmetic operations. The huge advances in NWP over the past few decades would have been quite impossible without simultaneous and truly remarkable increases in computing speed and storage capacity.

The primary weather elements used by NWP models are: wind speed and direction, temperature and humidity. From these, 'internal' elements are generated and carried through the forecast, they include: cloud type, percentage cloud cover, water and ice content, vertical wind speed and ground state (e.g. whether wet, dry or snow covered). In addition, for each grid point there are many fixed factors, amongst them: latitude; longitude (and hence incoming and outgoing radiation); whether sea or land; height of land above mean sea level, sea temperature; soil or vegetation type and cover. NWP models are now almost invariably combined with sea wave models to give forecasts of sea conditions as well as weather.

3 Retrieving the results

Completion of the NWP model forecast run is not the full story, because at that stage the results are simply a mass of numbers inside a computer. The forecast must be output in a user-friendly form, and often in many different forms, depending on the uses to which it will be put.

Airlines use forecast upper winds and temperatures for flight planning. Since these plans are themselves largely computer derived, grid point forecasts of upper winds and temperatures are transferred directly from the meteorological NWP computer to that of the customer.

For the multiplicity of personal and business users there are many ways in which the forecast data can be presented. Meteorologists require forecast charts in conventional format, with isobars, frontal zones and expected cloud and rain areas (see Figure 7.1). These are also studied on VDU screens where movie sequences can be revealing. Newspaper and especially television presentations have charts set up using specially designed graphics to show expected wind, cloud, temperature and weather variations over different time periods. Animated sequences are becoming increasingly sophisticated and popular for television weather forecasts.

Shipping interests may now receive wind and wave forecasts by facsimile (fax) directly from computer predictions in the form of symbols or plain language. Text broadcasts and warnings still usually require the intervention of trained forecasters.

4 Interpretation

The skilled and experienced weather forecaster still has an important role. Local effects on weather are inadequately predicted by NWP models with their limited resolution and data. An experienced forecaster can add detail to the forecasts, and by monitoring the weather he or she may adjust the output from NWP models in respect of the locality.

NWP wind forecasts through narrow straits such as the English Channel provide less detail than an experienced forecaster can produce. He or she will use NWP products for guidance on large scale changes, but 'tune' them for local effects. Indeed, small-scale geography has a considerable influence on wind and temperature in many places.

The likelihood of showers at or near a grid point can be indicated by a NWP model by means of the forecast stability and humidity profile. However, the distribution in space and time of shower development is highly dependent on local topography and a forecaster will often add value. Furthermore mist, fog very low cloud are often shallow and localised; detail is not handled well currently by NWP models.

Model Output Statistics

An important statistical technique has been developed to add local detail to specific forecasts from NWP models. For example, the day maximum temperature at a particular town may be influenced strongly by a lake in one direction and small hills in another, with the NWP model quite unable to resolve them. The Model Output Statistics or MOS technique compares day-by-day variations of the maximum temperature over a long period to all the various items forecast on each occasion by the model at nearby grid points. Using a technique called statistical regression, the best relationships are found and expressed mathematically. These are used in the computer for future forecasts, and are updated in the light of experience. In the case of the maximum temperature quoted above, it may be that a slight difference in low-level wind direction brings air across the lake and reduces it in summer but increases it in winter; MOS takes this into account. MOS enables useful forecasts of items that would normally lie well outside the NWP model's capacity. What is more it can be applied to non-meteorological variables – such as the likely sales of ice cream, gas consumption, infestations of aphids, likely numbers of hay fever sufferers, and many more.

Predictability and chaos

Meteorologists obtain worldwide pictures from satellites, accurate observations from aircraft, data from drifting buoys in the nearly empty oceans and even rain areas from radar networks. Sophisticated NWP models of the atmosphere run on computers, the power of which would until recently have been unthinkable, and produce forecasts of a quality thought impossible only a few years ago. There is clearly room for further progress, but how much is possible? Is there a limit beyond which the atmosphere is unpredictable? When will the increased resources required to advance further produce diminishing returns? To explore this concept of predictability let us first consider two simple mechanical systems.

When a ball is released on to a rotating roulette wheel it bounces wildly before eventually settling in one sector as the wheel slows to rest. The result is completely unpredictable even though the mathematics and mechanics of rotating systems, bouncing balls and trajectories through the air are all well known. This is because minute differences, far smaller than can be measured, have an overwhelming influence on the result. A variation in the point of impact of the ball much less than the width of a human hair, changes its trajectory sufficiently to alter the next point of impact by many centimetres and completely alter its progress thereafter. The result is RANDOM. We know the atmosphere is not like a roulette wheel; it is coherent and often predictable on the large scale for several days ahead.

Now consider a rotating smooth saucer, where a ball is released initially from near the rim and rolls towards the centre. It will clearly overshoot and be carried by its impetus up the other side. This oscillation will recur many times, slowly diminishing as friction brings it gradually to a halt. It is almost entirely PREDICTABLE. It can be repeated with similar results, and small differences in initial position play a minor role in the end result. The atmosphere is not like this; it sometimes changes dramatically over a wide area in the space of one or two days.

It appears that the atmosphere is a combination of both these types of system. Mid-latitude depressions often develop, move and decay in a regular if not identical manner for several days or weeks, unaffected by smaller scale disturbances, much as the ball in the saucer. Then the pattern of upper winds subtly changes, and suddenly the track they take becomes quite different, or they become stationary. Areas until recently subjected to days of wind and rain with little respite, now lie under the clear skies of an anticyclone for weeks on end. Modern NWP models of the atmosphere now sometimes predict these major changes of circulation type (see also the index cycle in Figure 4.5) up to a week ahead; on other occasions changes are not well predicted with the most powerful models in the world disagreeing amongst themselves.

Scientists have long been aware of inherent limits in our forecasting ability. These arise from uncertainties in our knowledge of the initial state of the atmosphere; from known small-scale detail impossible to include in our relatively coarse NWP models; and from inaccuracies in the solution of the complex equations. It now seems certain that the atmosphere itself does not have a unique path of evolution. To put it another way, the atmosphere itself may be 'undecided' what will be happening in two or three weeks' time. It has been said graphically that our future weather may depend on the 'flap of a butterfly's wings'!

A system which appears to be regular and predictable but which suddenly changes its behaviour to become completely and consistently different, is called **chaotic**. Imagine our ball on the saucer as described above, but with the saucer gradually tilting. Initially the regular motion of the ball proceeds much as before, enabling its future path to be predicted with some confidence. However, the point is reached when the ball suddenly falls to the floor, its earlier motion is at once irrelevant and predictions based on it will be seen to be quite wrong.

Theoretical and laboratory studies have found that for a planet the size of the Earth with its gravitational force and speed of rotation, it is impossible for our atmosphere to reach a steady state, or even a regular oscillation. Truly we have a restless atmosphere. This is indeed fortunate, because if depressions always took the same track then areas affected would soon become waterlogged, while elsewhere deserts would span the globe.

Cycles and forecasting

There are many regular features of our weather machine; the Earth's annual elliptical orbit round the sun, and its daily rotation are prime examples. However even these vary slightly. The Earth's orbit gradually changes shape over a period of about 95 000 years; its axis wobbles slightly over about a 23 000-year cycle, and its tilt in relation to the ecliptic (the plane in which the Earth orbits) also varies through a 41 000-year cycle. None of these is likely to worry us in our lifetimes, but each is believed to contribute to the onset of **ice ages**, long periods during which polar ice extends to 50 degrees latitude or less. These long period cycles are called **Milankovitch variations** after the scientist who first postulated their importance.

There are also much shorter periodic variations which it has been claimed influence our weather, the favourite being the sunspot cycle. A **sunspot** is a dark, cooler area on the sun's surface, marking a major eruption or storm from its interior which may last from a few hours to weeks. While they may occur at any time, there is a consistent and often marked peak in sunspot activity every 11 years (approximately) that has become known as the **sunspot cycle**. Activity of the sun, in the form of output of cosmic rays, increases with the presence of sunspots; this affects the Earth's ionosphere, and hence radio transmission.

In the stratosphere near the equator there is a well marked, but irregular, oscillation in the westerly winds, lasting about 27 months on average. This is called the **quasi-biennial oscillation** or QBO, and is explained by the effects of weather systems in the equatorial troposphere upon much longer seasonal changes in the stratosphere.

The **El Niño southern oscillation** or ENSO is an extensive warming of the eastern tropical Pacific Ocean which occurs every few years, and lasts for several months. It is not regular, and affects the weather, especially rainfall over Ecuador and Peru. But perhaps more important is the marked depletion of fish in the warm currents with serious knock-on effects on the economies of those parts of South America.

Tides, due to the gravitational pull of the moon and sun, are a well known and visible feature of the large oceans. Less well known are similar unseen effects in the atmosphere where the sun is the major influence. The major part of the sun's 'tidal' effect is due to the contrast in heating between the daytime and night-time hemispheres. There is a regular atmospheric pressure oscillation through a 12-hour cycle, varying from about 4.0 mb at the equator to near zero at the poles. It has a value of around 1.5 mb at middle latitudes with the maximum occurring at near 10:00 and 22:00 local time. Forecasters need to be aware of this to avoid misinterpreting pressure changes on their charts.

Apart from obvious local effects of the ENSO the above cycles make no useful contribution to weather forecasting. That is not to say they have no effect on our weather – they must do, but that it is individually and collectively trivial. For example suppose the variability of weather in California and the United Kingdom could be represented by cakes – the former would be smaller but arguably more tasty! The contribution from the sunspot cycle might be a small crumb and that from tidal effects another. The QBO and ENSO would contribute slightly larger crumbs. It seems quite clear from years of research, that the total effects of all these cycles would not add up to the smallest possible slice, and are quite insignificant.

Ensemble forecasting

Because the atmosphere is a chaotic system, there are times when NWP models successfully forecast for more than a week ahead and others where even the next two or three days are subject to considerable uncertainty. For some years forecasters have tentatively assessed the likely reliability of a forecast by studying the consistency or otherwise of outcomes from successive runs of a model, and by studying results from different models covering the same period. This has not worked particularly well.

Increased computing power has made it possible to run a simplified NWP model over and over again for a forecast period of up to ten days, each time slightly changing the initialisation. The collection of forecasts obtained in this way is called an **ensemble**. When members of the ensemble are all broadly similar to the main NWP model run, then it appears that the predicted evolution is not sensitive to initial conditions, and is unlikely to be upset by minor unforeseen influences (or a butterfly's wings!); confidence is high. On the other hand, when the ensemble of forecasts diverge widely through the forecast period, then the atmosphere appears to be unpredictable and confidence is low. The ensemble technique is in its infancy, and continues to be investigated and refined.

Probability forecasting

Forecasts can sometimes be expressed in terms of the likelihood of certain outcomes, for example the probability of overnight fog or frost at a place, or of rain interrupting an outdoor event. This is probability forecasting. The ensemble approach seems ideally suited to probability forecasting because often the total ensemble splits into several sets or clusters each with similar outcomes. This may sometimes enable forecasters to predict a week ahead that, for instance, warm weather is twice as likely as cold weather. Even this vague forecast may be of value; for example if a customer needs to shut down a low-cost power station for maintenance, he or she will wish to maximise the chance of doing so when demand is below average and a high cost alternative is less likely to be needed.

Long-range forecasting

Long-range forecasts are generally regarded as covering the period beyond a week or so ahead. They are usually restricted to summarising the main character of the weather expected over a month or season, relative to average climatological conditions over previous years. The desirability of long-range forecasting has rarely been questioned, especially where weather is variable such as the British Isles. However, the ability to forecast accurately many months ahead, fair and foul summer weeks, or even good growing seasons for one crop compared with another, would not be an unallayed blessing. Tourism would be hard hit and agricultural output distorted.

In the past, long-range forecasting in the United Kingdom and elsewhere was based largely on statistical comparisons between recent averaged weather patterns in terms of temperature and pressure, and those of previous years. The theory was that if over one month there was a good parallel with that month in one or more previous years, then the evolution should also be similar over the next month or season. Perhaps through a lack of historical data, but most likely a completely false hypothesis, such forecasts were poor and have been long discontinued.

Many operators have claimed to have devised methods enabling accurate detailed forecasts to be made weeks or even months ahead. None has (yet) stood the test of time.

It is now considered that NWP models offer the best prospect, especially where the effects of ocean currents are included within them. There are large and sometimes persistent variations in sea surface temperature from the seasonal average over wide tracts of the oceans. These **sea temperature anomalies**, as they are known, affect the heating of overlying air and the evaporation of water into it. They may influence rainfall over upwind continents in the following season, but effects are likely to be more widespread and complex. NWP models have an important part to play, not only in predicting the weather changes resulting from such anomalous temperatures, but also in investigating the mechanisms. Research continues.

Future advances in forecasting

Over the past several decades, advances in weather forecasting have come from three mutually supportive directions: observations, understanding and powerful computers, each drawing on international cooperation and scientific endeavour of the highest quality. Future advances are likely to also be in these areas.

Observations are essential. Satellites, radar, much improved aircraft observations together with drifting buoys have compensated for the loss of expensive-to-run weather ships. With future higher resolution models it will be even more important to increase the amount and geographical distribution of high-grade information. Furthermore, it will be necessary to continually monitor the effect of low-grade satellite data such as satems and satobs (see 'remote sensing', Chapter 6) on NWP forecasts. Short-period NWP forecasts are already so good that some data may degrade, rather than add detail to, initialised fields for the next forecast run.

Understanding is vital. In the early years of NWP models it was often found necessary to reduce consistent errors by incorporating non-meteorological factors in the mathematics. This inevitably led to difficulties when subsequent refinements were introduced. Research into many aspects of meteorology continues and will remain necessary to enable us to refine NWP models and improve their accuracy.

Computing power will need to continue its remarkable increase with time. There is every indication that it will.

—— Interpreting a weather chart ——

Usually a weather chart in a newspaper or requested by dial-up fax will give a fixed time picture, including the position and intensity of weather systems and fronts together with isobars. A great deal of information may be gleaned from this, especially where it can be combined with a professionally prepared forecast. The Internet is another source of plentiful weather information, including forecast charts, satellite pictures and scripts.

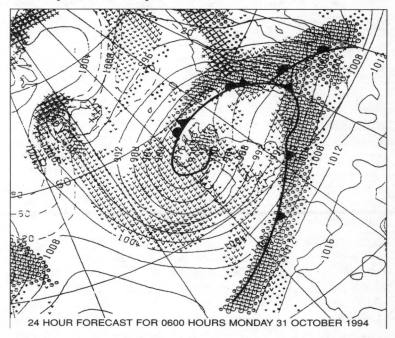

24 HOUR FORECAST FOR 0600 HOURS MONDAY 31 OCTOBER 1994

Crown copyright

Figure 7.1 A 24-hour forecast chart for NW Europe from the UK Met Office Limited Area Model. Fronts have been added to the isobars and precipitation symbols. Round symbols denote rain at that grid point, X's snow and V's showers; in all cases the larger they are the heavier the precipitation forecast. A mature occluded low lies west of Scotland with the remains of its warm sector sweeping into southern Sweden. There are broad areas of frontal rain, and showers near the low and in the cold northerly to the west of it. Over the high ground of southern Norway snow is forecast, and also north-west of Iceland. In the high north of Iceland and the ridge west of the showery northerly there is no precipitation.

It is not difficult to associate features on weather charts with broad-scale weather, and to focus on the local area. However, a professional forecaster has the advantage of knowing what is happening above the surface, too, and how the weather picture is likely to evolve in time. It is never wrong to seek expert advice.

Weather systems were discussed in Chapter 5 and the formation of rain in Chapter 3. Here we will look at a few rules, and apply them to actual examples typical of weather situations over the United Kingdom. The same reasoning may be applied anywhere on the globe, but be aware of differences due to topography (especially sea and mountains) and because of latitude. If necessary, consult the 'Climates of the World' in Chapter 12.

Rules for the British Isles

Wind

Wind blows 30 degrees or so across the isobars with low pressure to the left. Wind speed is strongest when isobars are closely spaced and light when they are wide apart. It is reduced inland compared with windward coasts (see sea breezes in Chapter 5), but is almost invariably stronger over hills and mountains, sometimes considerably so.

Temperature

Northerlies are almost always cold, and southerlies warm, relative to the seasonal average. Assess where the main windflow is coming from and refer to 'air masses' in Chapter 5. A long sea track normally means a westerly or south-westerly, and is mild but usually cloudy in winter, and cool in summer. A long land track bringing easterly or north-easterly winds is hot in summer but bitterly cold in winter. High pressure brings light winds which means cold weather in winter, and cold nights with frost in spring and autumn; but in summer produces the hottest weather especially if it is centred to the north or north-east. Temperature extremes are never associated with strong winds.

Rain (or snow)

Continuous precipitation, lasting an hour or more is most often associated with depressions and weather fronts which are discussed in

Chapter 5. Normally fronts are most active where pressure is low (say below 1000 mb) or falling; they are usually weak often with little or no rain where pressure is high (say above 1020 mb) or rising.

Serious snow most often results from a warm front or occlusion pushing in slowly from the west or south against well-established cold winds. However, polar lows moving south can produce large amounts in the north, and winter depressions running east through the English Channel give copious snow over southern England and Wales. It is difficult even for a professional forecaster to predict exactly where the boundary between rain and snow will be when this happens. Treat newspaper forecasts with great caution, they were prepared many hours previously.

Showers

Showers are always likely in a north-westerly when pressure is low, and also on and to the rear of cold fronts. They are always enhanced if isobars are cyclonically curved (pressure is lower within the curve) and also when pressure is falling. Showers are inhibited when isobars are anticyclonically curved and also when the pressure is rising. Often high surface temperatures (relatively) are required to trigger their parent cumulus or cumulonimbus clouds. This means that inland they are much more frequent from late morning through the afternoon, and are generally less common in winter. Winds off the sea bring showers by night and day to coastal areas when seas are warm and the airstream cool; that usually means autumn and winter. Hilly areas trigger more showers than flat terrain.

Thunderstorms

These occur when shower clouds are particularly deep, and may be regarded simply as a type of heavy shower (but see Chapter 10). The most widespread and most violent thunderstorms occur in summer when hot and humid low-level air is overridden by cooler air aloft ahead of fronts, usually with falling pressure (see the discussion on 'tornadoes' in Chapter 10).

Fog

Fog forms on radiation nights when skies clear and winds are light, which happens most often in anticyclones or ridges of high pressure.

Long winter nights can produce freezing fog which the weak sun may not manage to clear during the day. Summer fog, on the other hand is short-lived and often the precursor of a hot day.

Drizzle

This often occurs in the warm sector between a warm front and following cold front, but is much more frequent over coasts and hills where warm moist air is forced to ascend, especially from the southwest. Persistent drizzle, low cloud and fog often occur in cold easterly winds across North East England and Eastern Scotland when it is termed haar or fret, especially when the sea is very cold in the spring and early summer. Sometimes overnight fog (see above) thickens enough to produce drizzle, a good indicator that the fog will be slow to clear.

8
WEATHER AND THE ECONOMY

Almost every aspect of our lives is affected by the weather: where we live, what we wear, eat and drink, and our work and leisure pursuits. It imposes costs on us but also bestows benefits. Storms and floods cause damage and sometimes loss of life, but the day-to-day variations contribute to the rich variety of vegetation and fauna, which add so much to the quality of life.

To derive a balance sheet or profit and loss account for weather is not possible, but we can be sure that forecasting plays a part in assisting the wider economy. Valuing weather forecasting is difficult, but an independent survey carried out for the UK Met Office in 1994 found that weather forecasts save Britain almost £1 billion annually, spread across the whole spectrum of commerce, industry, government and, not least, the general public. Here we look a little closer at the effects of weather on various areas of economic and social activity, and likely benefits from forecasting.

Agriculture

Farming remains fundamental to our survival, and the effects of weather are far-reaching. First, the crops which can be grown in any locality depend on the climate: the amounts of sun and rain, temperature range and the length of the growing season. Rearing animals relies on the availability of grazing or fodder and a comfortable climate through most of the year. Animals distressed through excessive heat or cold, quite apart from the cruelty involved, will not be productive. While livestock can live indoors for part or all of the time, just as plants may be provided with an artificial environment under glass, these provisions involve extra costs which must be taken into account. In today's world restrictions on international trade are likely to continue to diminish. That means low-cost agriculture in suitable climatic conditions, even adding transport costs, may make production of many crops elsewhere in artificially maintained climates uneconomical. Few plants grow successfully and are economically viable further north (or south) than about 60 degrees latitude, which includes much of Canada, Scandinavia and Russia.

Climate is of prime importance, but day-to-day weather variations can be crucial. Citrus fruit trees are sensitive to frost, as are many other fruit blossoms and coffee bushes. Forecasts of late frosts allow preventive measures to be taken; for example, spraying with water to prevent a large overnight temperature fall. Local climatic variations, or the microclimate, can be equally important. Soft fruits thrive on south-facing slopes in Northern Europe, whereas they would be unproductive on shaded land much further south. It is easy to overlook the humble crop of grass, but in many economies such as those of New Zealand, Argentina and Ireland it is of great importance. A necessary climatic variable for grass to be successful is summer rain, and that makes Mediterranean and tropical climates ill-suited to produce pastoral products such as milk, butter, meat, cheese or wool.

Droughts are always a threat to crops, not only in temperate regions. They may be countered by irrigation or, over a longer perspective, by developing drought-resistant varieties of popular crops. In the early 1990s much of southern Africa was seriously affected and predominantly agricultural economies were brought to their knees. Floods are also a hazard for agriculture, except where they are part of the normal climatic variation. In the monsoon climates of Asia, the staple rice crop requires copious moisture and is usually grown in naturally or artificially flooded paddy fields.

Most pests and diseases are weather dependent. Locusts are probably the most voracious and have a peculiar life cycle. Their eggs lie inert in desert sands for many years before being activated by a rare shower of rain. They then hatch in their millions with the resulting swarm consuming every green shoot in their path. Weather forecasts not only indicate when hatching is likely, but also give the probable route of locust swarms to enable countermeasures such as spraying to be most effective. Aphid and fungal infestations such as potato blight, become severe only when the temperature and humidity exceed certain thresholds. Forecasts of these occasions enable growers to anticipate events and spray when it is likely to be most effective.

Livestock are particularly subject to stress at certain periods, not only when new-born. For example, sheep are especially prone to disease and exposure after shearing if the weather is cold and wet. Farmers can be advised to either adjust their programme to avoid wet and windy spells, or keep sheep in shelter for a few days after being shorn.

In high latitudes, and on mountain slopes elsewhere, temperatures often fall well below freezing and high winds are commonplace. In this harsh environment pine forests thrive where little else will. These provide shelter and fodder for indigenous animals such as reindeer, and also comprise an important renewable source of soft wood and pulp. At the other extreme tropical rainforests produce most of the world's hardwoods, which thrive in a hot and humid atmosphere. These fast growing forests absorb vast quantities of atmospheric carbon dioxide (CO_2), and the present trend to forest clearance has contributed to the increase of CO_2 in the air (see Chapter 11).

Fishing

Fishing is highly weather dependent because small vessels cannot operate safely in stormy weather. Shipping forecasts concentrate on wind speeds, because even if high winds do not threaten safety, they may make it impossible to launch and retrieve nets. Early warnings of gales or storms enable fishing fleets to take shelter, and outlooks for several days ahead enable the selection of those fishing grounds offering the best conditions. From this, it can be seen that weather forecasting and the fishing catch can be closely related.

There are more serious aspects. Many of the most productive fishing grounds are in high latitudes, for example in the North Atlantic near Iceland, and in the South Atlantic near South Georgia and the South Sandwich Islands. In these regions air temperatures can fall well below freezing and, in strong winds, sea spray will freeze, building up ice on cables and superstructure. Since the capsizing of several trawlers in Icelandic waters from severe icing in the 1960s, warnings have been issued of these conditions to enable vessels to make for port.

Manufacturing

The siting of a factory depends on many factors, of which weather and climate may appear well down the list. The availability and cost of land, labour, power, water, transport facilities and the investment climate are the main variables, but of course many of these are weather dependent. A manufacturing plant inaccessible for much of the year due to waterways being frozen over may not be viable, even if the land upon which it is built is cheap. An airport built on fog prone marshland may not appeal to passenger or airline customers. Planning and architecture always need to take into account heating and air conditioning requirements. In general, buildings are designed to provide their own microclimates, and day-to-day weather dependence is limited.

Construction

Few aspects of construction are not weather dependent, and many are greatly so. Concrete laying and pouring, and the subsequent curing are susceptible to frost and rain, with a substantial cost penalty if large volumes are affected and have to be re-layed. External work is generally sensitive to low temperature, rain and strong winds, with productivity falling off markedly, especially where cranes or scaffolding work is involved. Groundwork such as trenching for foundations, landscaping and paving cannot take place if the ground is frozen or waterlogged.

The planning of every construction project needs to take weather into account, from the earliest design and tendering stage through to completion. Initial design will be strongly influenced by climate in deciding the degree of insulation, heating and air conditioning required. Wind statistics will be used to determine the strength needed by long and exposed bridges and towers. The tender must be adjusted to allow for weather risks and the possible impact on construction cost. The risks are usually computed from a statistical analysis of past records from near the locality.

During the construction phase itself weather forecasts enable the site manager to plan the programme of work to minimise interruptions due to inclement weather. Labour may be switched from exterior to interior work and vice versa. Sometimes two or three successive fine days (a so-called **weather window**) are essential for a particularly susceptible stage of construction, when the weather forecaster has an important part to play. This is particularly true of huge oil production platforms which are constructed inshore and then floated to their installation site at sea.

The 1994 UK study into the benefits of weather forecasting estimated that in Britain the construction industry saves over £100 million annually by heeding weather forecasts. There is little doubt that similar proportional savings are made in most developed countries.

Transport

All sectors of the transport industry are affected by weather, and it has been estimated that weather forecasts enabled the UK transport industry to save over £250 million in 1994. Adverse unexpected weather conditions cause delays and accidents, and hinder rescue and recovery operations.

Airlines

Forecast high-level winds and temperatures are always used in flight planning. They enable the benefit of tailwind components to be harnessed where possible and the worst head-winds to be avoided. Furthermore, the fuel load may be reduced in favourable circumstances increasing efficiency. Jet engines, especially on supersonic aircraft, run more efficiently at low temperatures, consequently forecast temperatures are also taken into account. Weather forecasts enable airlines worldwide to save more than $1.0 billion every year in fuel.

Aircraft have always faced weather-related hazards in flight, although the design of modern large passenger jets has made them much less susceptible. Nevertheless, forecasts are issued as routine across the globe allowing avoidance or at least precautions to be taken, such as advising seat belts to be used.

- **Clear air turbulence (CAT)** occurs outside of cloud and is, of course, quite invisible. It is due to rapid changes in wind velocity and causes aircraft to be shaken, sometimes violently. It is encountered where jetstreams are particularly strong, or abruptly change direction. It also results from strong winds being deflected as they cross hills or mountains, the effects sometimes being felt many thousands of feet above and hundreds of kilometres beyond, and also near cumulonimbus clouds.

- **Airframe** or **engine icing** occurs when cloud droplets (or raindrops) below freezing immediately turn to ice on impact with a cold surface (see 'rime' in Chapter 6). Large build-up on leading edges of wings will affect lift, and in air intakes can cause engines to lose power. De-icing equipment on big modern aircraft soon removes the hazard, but smaller fixed and rotary wing aircraft may be more susceptible and should always heed warnings. Severe icing occurs only when the water content of the cloud is particularly high; this usually involves nimbostratus or cumulonimbus. Icing is especially severe on the rare occasions when rain falls through air whose temperature is below freezing. This is known as **rain ice** and produces rapid ice accretion; it should be avoided by all aircraft.

Landing and take-off may be made impossible by fog or low cloud, and at some airfields by strong winds blowing across the runway, or **cross winds**. Forecasters prepare **Terminal Area Forecasts**, or **TAF**s, for every major airfield and most minor airfields all over the world, covering up to 24 hours ahead. These are regularly updated and made available to air traffic control and aircraft in flight. Flight delays or especially diversion to other than the destination airfield, is costly and inconvenient for operators and passengers alike. Accurate forecasts help to avoid this, but automatic take-off and landing systems at major airfields now enable suitably equipped aircraft to operate even in dense fog. There remains, however, the problem of manoeuvring on the ground!

Fog dispersal has been tried with some success. In the United States where air temperatures in winter fog are often well below freezing, dry ice and silver iodide crystals have been found to cause many of the fog droplets, which are supercooled water, to 'freeze out' and fall to the ground, considerably improving visibility. At military airfields in southern England during the Second World War powerful burners were used to thin fog along runways by evaporation. Without doubt the greatest advance in fog reduction in the United Kingdom came with the Clean Air Acts of the 1950s, which progressively banned the use of other than smokeless fuels. In the London area the frequency of dense fog more than halved between 1950 and 1970.

Shipping

Modern ocean-going shipping is generally able to withstand the worst that the weather can throw at it. Nevertheless, there is an economic penalty as well as discomfort in encountering storms: cargo may be damaged, excessive fuel used and, not least, fare-paying passengers discouraged from repeating the experience. To be of value to shipping operators, wind and wave forecasts need to be reliable for two or three days ahead, which has become the case only in recent years. Storms, especially destructive hurricanes, need to be pinpointed and their movement forecast accurately to enable avoiding action to be taken.

General 24-hour forecasts, including warnings of gales, are issued for all seas and oceans by appropriate national meteorological services, under the auspices of the World Meteorological Organization (WMO). In addition, many operators of ocean-going fleets take a bespoke **weather routing** service for each of their vessels. Every voyage is closely monitored by a team comprised of forecasters and master-mariners. They determine the optimum route using computer-generated forecasts of wind and waves, paying due regard to the vessel and cargo concerned. Forecast updates and advised route adjustments will be sent by radio at least once a day. A route might be required to maximise passenger comfort or minimise cargo damage by avoiding high seas, but perhaps most commonly will minimise fuel used and time taken. While savings are difficult to measure, and for some voyages are negligible, there is no doubt that weather routing of ships is highly cost effective, especially with the improved forecasts of recent years. With the running cost of many vessels exceeding $1000 an hour, which is the approximate total cost per voyage of weather routing, an average saving of half a day per voyage represents a considerable return on investment, irrespective of environmental benefits.

Coastal shipping is by no means independent of the weather. Not all ferry tragedies in European waters in the 1990s have been weather related, but tropical storms have certainly led to disasters close inshore in the Philippines and Bangladesh. These should be avoidable with timely weather warnings. When winds reach storm force (see Chapter 6) ferry services in the English Channel are normally suspended.

Ocean tides are driven by the gravitational pull of the moon as it circles the globe. They are independent of the weather, but may be dangerously reinforced by wind-driven **storm surges**. These occur when strong and persistent winds pile water up over and above high tide level, and are highest where winds blow across a closed basin such as the North Sea. Widespread flooding and tragic loss of life in Holland and eastern England in 1953 was the result of a particularly high tide coinciding with strong winds, as a deep depression crossed the northern North Sea. Following this, sea defences were strengthened and a moveable barrage called the Thames Barrier built across the Thames estuary. In addition, much research was carried out into storm surges and mathematical models developed to forecast the day-to-day effect of expected winds on tide levels. The United Kingdom Storm Tide Warning Service (STWS), operated by the United Kingdom Met Office, predicts and issues warnings of unusual tide levels. Sometimes negative surges occur, when water levels at low tide are further decreased by winds taking water out of the basin. It is important to forecast these too, because large ships with deep draught may be temporarily unable to navigate the usual routes through the English Channel without the risk of grounding.

Road transport

Safety is the major consideration when effects of weather on road transport are considered. Fog, snow, icy roads, strong and gusty winds, and heavy rain all cause accidents with consequential injury and loss of life, quite apart from the economic penalty of disruption. Heavy and drifting snow, and dense fog are particularly hazardous and can bring traffic to a halt. Weather forecasts highlight possible problems, enabling motorists and transport managers to switch routes or to reschedule their journeys, undoubtedly preventing accidents and a good deal of frustration. Large high-sided trucks are particularly at risk of being blown over by strong and gusty winds, especially on open stretches of road. High bridges are closed to high-sided vehicles (sometimes all vehicles) when winds are blowing across the bridge and exceed gale force. There is a considerable economic penalty in closing busy routes and, while accurate observations minimise the period of closure, timely forecasts enable alternative routes to be used to reduce delays.

In a variable climate such as that of Britain, icy roads provide a more frequent if transient problem than in even much colder countries. The dangers can largely be overcome by spreading grit or salt on road surfaces: grit helps tyres to grip, while salt lowers the freezing point of water and prevents ice formation. Both measures are expensive and to some extent polluting. If road surfaces are likely to stay above freezing, or are dry and likely to remain so, or if the wind will dry wet roads before ice forms, then treating the roads is not worthwhile. Furthermore, if salt will be washed away by rain before ice can form, spreading must be delayed.

A great deal of research has been undertaken into forecasting ice formation on roads. It was found that not only the local geography of the road but also the subsoil under it were important. For this reason, detailed measurements are made of how the road temperature behaves at a representative selection of sites; this is called **thermal mapping**. Subsequently observed temperatures at these sites from remote-reading thermometers are related each day to the forecast minimum temperatures in order to determine the likelihood of ice formation. By this means, forecasters are able to produce site specific forecasts, allowing salt or grit spreading only where ice formation is most likely, rather than an all or nothing approach. Savings over a whole winter season can exceed one hundred times the cost of this service.

Rail

Rail transport is less susceptible to bad weather than road traffic, nevertheless smooth running depends on early warning of likely problems. For example: snow warnings enable plough units to be fitted to engines, or special snow and ice clearance trains to be run; strong early autumn winds bring falling leaves which insulate electrical conductor rails and must be swept clear before they cause trouble; warnings of frost enable antifreeze to be applied to railway points which otherwise can freeze and become inoperable. Trains are rarely affected by heavy rain, unless it results in flooding or subsidence, and fog is much less of a problem than for road or air transport. Extremely fine, penetrative snow can occur in exceptionally cold weather (see Chapter 6). In England in the early 1990s this caused serious problems to the latest locomotives, and in consequence popularly became known as 'the wrong kind of snow'.

Tall railway wagons, such as those of Eurotunnel's *Le Shuttle* service which are 5.6 m (18 ft) high, may be adversely affected by strong winds when they are out in the open. To counter this, certain of the tracks at the Channel Tunnel terminals are sheltered by specially constructed 'wind fences'. When high gusty winds are forecast the wagons are restricted to these protected tracks.

Power utilities

There is a strong and obvious relationship between weather and fuel used for heating (and cooling). The most important factor is air temperature, but wind, humidity, rain or snow and even cloud amount can all be relevant. Where individual buildings or blocks have their own local fuel supply, such as a tank of oil or bunker of coal, then weather fluctuations producing a surge in fuel demand can be met. On the other hand, electricity and gas networks must be geared to satisfy large and often sudden increases in demand across their whole network.

Electricity and gas authorities have long realised the benefits to their businesses from accurate weather forecasts. Careful studies are carried out into the effects of past weather variations on power consumption. From these, it is possible to anticipate changes in demand using forecasts of the various weather elements over periods ranging from the next hour or two to several days. The onset of a predicted cold snap will then find electricity authorities with the necessary extra generating plant ready and waiting; gas companies will have forewarned clients with interuptible supplies that they may be disconnected, and will have compressors primed to pump extra gas into the grid. On the other hand, unexpected sudden increases in electricity demand will need to be satisfied by using fast-response but high-cost generating plant, such as gas turbines, or by voltage reductions; unexpected sudden gas demand will decrease pressure and the efficiency of consumers' appliances.

All power generating equipment needs periodical servicing and repair. Where it is not practicable to delay this until summer, weather forecasts may be vital to select periods when demand is likely to be low.

Weather plays an important role in generating electrical energy. Hydroelectric power is the modern equivalent of a watermill, using water stored in reservoirs or lakes at high levels to drive turbines as the water flows down the mountain side. Power output can be considerable and is cheap once the generating plant is installed. Furthermore, it is environmentally friendly, and can be adjusted by regulating the flow. Climatological studies are essential before the large capital investment is made, to ensure that high and reliable rainfall occurs over the catchment area. Mountainous countries such as Norway generate virtually all of their electricity in this way.

Windmills as well as watermills were used for centuries, usually to grind grain into flour. Modern **wind turbine** generators are much larger, and aerodynamically designed to extract maximum power from the wind for conversion into electricity. They need to be mounted high up or offshore to be exposed to the strongest winds, with hill tops and ridges the most favoured sites. Unfortunately, a single large wind turbine produces only about one megawatt compared with a few hundreds from even a small modern coal, oil or gas fired generator. (A megawatt is a thousand kilowatts, that is sufficient to run 1000 one-bar electric fires, or toasters.) Furthermore, wind turbines are rather noisy, require regular maintenance, expensive cable links to the nearest town or electric grid and are regarded by many as ugly. Nevertheless, with the political will it is possible to obtain economies of scale by building wind 'farms' comprising hundreds of turbines sited away from inhabited areas, usefully contributing 'green' energy. For the best returns it is essential to study the meteorology of potential sites, bearing in mind that wind turbines produce nothing in winds below about 8 kn (10 mph).

Wave generators use the colossal energy of sea waves, often the result of storms far away, to drive pumps or turbines. Progress has been slow because of the many difficulties: salt water corrosion, electricity links to them, maintenance and design being among them. Groups of wave generators in the open sea would present a serious navigation hazard, while inshore wave generators would be easier to install but produce less energy.

Even deep coal mining is not always immune to weather! When atmospheric pressure is low, inflammable and sometimes poisonous gases in the rock strata deep in the mines tend to leak out, and provide an additional hazard for miners. Weather forecasters can warn of this possibility well in advance, enabling more frequent tests for gas to be made in affected pits and, where possible, ventilation to be increased.

Communications

Over the next few years the use of overhead telegraph wires may become consigned to history, as more and more independent signals become packed into narrow microwave bands and beamed to and from our homes via satellites. Until then many of us rely on telegraph lines which are susceptible to strong winds, snow and ice. Strong winds frequently snap lines and bring down supports. Less commonly a great deal of damage may be caused by ice building up in cold weather. This occurs when the lines are below freezing and in fog or cloud for a prolonged period. Droplets of supercooled water are deposited in the form of rime ice, which in windy conditions builds up to many centimetres thick. More rapid but much less common icing occurs when rain or drizzle falls on to cold lines and freezes. In either case, the weight of ice can bring down power lines over a wide area.

Microwave transmissions, and those associated with television and radio are sometimes affected by atmospheric conditions. Thunderstorms with their intense electrical discharges cause interference. Occasionally rapid changes of temperature and humidity with height, cause radio waves to be refracted into the ground (see anomalous propagation of radar in Chapter 6), leading to a weak signal further afield and interference. This occurs most frequently in long-lived anticyclones.

Water

A **drought** is a prolonged period with much less rainfall than average but there is no absolute definition. Use of the term clearly implies stress on indigenous vegetation and animal life. Almost every year some part of the world is affected, and not only regions with low average rainfall.

Rising populations and living standards increase water consumption, while tourism brings heavy demand to many regions in their dry season. Even in the United Kingdom, picturesque wet areas such as Cornwall and Devon find it difficult to store enough water to last through the driest summers.

The meteorology of climate and rainfall is well known, and statistics exist for most areas of the world from which probabilities of long dry periods can be calculated. This might be, for example, the duration of drought probable on average once in 100 years. Provision of sufficient water then becomes an economical and political decision, balancing resources against priorities. Even in the middle latitudes of changeable westerlies the weather pattern sometimes settles down to produce long, dry, anticyclonic spells. Although long-range forecasting is not possible, meteorologists can give shorter term information on the likely need for irrigation, and longer term advice on the necessity or otherwise of making provision.

Management of waterways, storage and drainage depends on short-term as well as long-term rainfall. Sudden intense storms can overwhelm local drainage and lead to what has become known as **flash flooding**. If a storm moves along the course of a river, especially when it is already high, then flooding may subsequently occur more widely lower down the valley. River authorities carefully monitor short-term rainfall forecasts based on latest radar and cloud imagery.

Business and retail

It is an unwise manager who ignores the effect of weather upon his or her business, because it plays such a large part in determining what we buy and when we buy it, and what we do and where we do it!

In spring the first fine, warm weekend beckons the green-fingered out of doors, and sales at crowded garden centres soar. Outdoor sportsmen and do-it-yourself enthusiasts rush to update equipment and buy tools and materials. Warm, sunny weather dramatically increases sales of salad stuffs, drinks and ice cream. On the other hand, cinemas, theatres and clubs suddenly lose some of their appeal, and 'special offers' may be needed to attract customers. Benefits from well-timed advertising campaigns are considerable and weather forecasters have an important role to play. The first cold snap of autumn is a different story. Local shops or those with covered car parking become more attractive; many car batteries are found wanting and the demand for antifreeze increases a thousand-fold in a matter of days.

Almost every trade, business and profession, from plumber to doctor, mechanic to pharmacist, is affected by the weather, but none more than the retailer. Almost all major department stores and supermarkets now use a meteorological consultancy service. This enables weather forecasts to be translated into likely sales patterns, based on detailed research into their particular stores' past sales/weather relationships. By this means the mix and amount of stock can be adjusted, and displays altered to maximise sales.

At any time of year, but especially in winter, bad weather can discourage shopping for anything other than essentials. Snow in particular keeps most 'optional' shoppers away. There is, however, an exception. A wet day may markedly increase retail trade in tourist resorts because outdoor attractions quickly lose appeal. This underlines the importance of studies into the effects of weather, rather than making assumptions.

Relations between retail trading and weather carry through to advertising and marketing. Special offers, advertising and sales campaigns may be geared to increase their effectiveness by anticipating weather-reinforced demand. The strategically aware distributor will ensure that the weather dependence of his or her product is used as an argument for building up stocks ahead of demand. Sales executives need to know how performance is connected with abnormal weather conditions, and be armed with appropriate statistics.

Commodities and futures

Every basic agricultural product depends on the weather. For producers to be sure of a minimum return, even in years of glut, it is customary for part of their anticipated crop to be sold in advance at a fixed price. These 'futures', as they are known, are marketed and may be bought and sold on commodity exchanges. Purchasers may be similarly placed; for example, a producer of potato crisps cannot budget effectively if the price of his or her major ingredient may double overnight. By purchasing 'future' contracts he or she can cover most requirements, hoping to make up the remainder more cheaply on the 'spot' or actual market when the crop is harvested.

The day-to-day price of futures varies, depending on demand in the light of the latest perceived likelihood of glut or shortfall in the eventual harvest. To a large extent this relates to weather through the growing season. If accurate long-range weather forecasting was possible, then a large part of this uncertainty would disappear. As it is, speculators gain some benefit from studying forecast weather on shorter timescales, especially where a single product is concentrated in a few territories.

A good example of this is the relationship between coffee futures and frosts in Brazil. About 25 per cent of the world's coffee is grown in Brazil, concentrated in latitudes between 25 and 35 degrees South. This region is susceptible to frost on rare occasions, when cold southerly winds fall light, under clear skies during the southern spring. Forecasts of these conditions enable large rises in the price of coffee futures to be anticipated.

Insurance

Many kinds of weather-related insurance are written each year, from the '**pluvious**' type of policy which insures a specific event against a rainy day, to standard household insurance which normally includes storm damage. Actuaries deciding premium levels need detailed meteorological advice on the likelihood of damaging weather, especially winds and floods which vary considerably from place to place. Hailstorms are particularly damaging to many crops, spoiling fruit and beating down cereals. In the United States farmers regularly insure against this risk. In Europe the risk is smaller except in the east; there the main method of insurance has been to attempt to diffuse hailstorms by firing cannon shells into storm clouds, but its efficacy is not proven. In areas where tornadoes and hurricanes are a high risk, storm damage is virtually impossible to insure against.

Marine and aviation insurance premiums similarly include an element covering risk of loss or damage from weather events, for which meteorological advice is needed. Conditions may be laid down in insurance policies; for example, ship operators plying routes in low latitudes in the hurricane season may be required to take professional ship routing advice to help avoid severe weather.

—— Leisure, sport and tourism ——

Holiday resorts depend greatly for their appeal to holidaymakers on their climate, and the wealth of many areas is built on tourists' high expectation of fine weather. For day-to-day leisure purposes and holidays off the beaten track, weather is often equally important. An unfavourable weather forecast deters many day trippers or weekend holidaymakers from going ahead, though operators of indoor entertainments benefit. Some areas have accessible and beautiful coasts facing opposite directions, and it is often possible for holidaymakers, sailors and surfers to choose the most suitable beach for them in the light of forecast winds.

Recent years have seen large indoor holiday parks being built in central and northern Europe. These have artificial beaches, bright lighting and wave machines, providing 'a summer's day at the seaside' whatever the weather and time of year.

Many outdoor leisure pursuits are highly weather-dependent and for safety it is vital that up-to-date weather information and forecasts should be heeded. Mountaineering, hill walking, skiing, gliding, sailing, sailboarding and sea fishing are obvious examples. Unfortunately, many lives are lost each year on small boats, even close inshore, and on hills and mountains when rapid weather changes catch people off guard. Sometimes the disappointment of calling off an expedition after weeks of planning is too great, and a forecast of bad weather is ignored when the immediate weather prospects look fair. Tragedy all too often follows. Caving or pot-holing underground is not immune. Heavy rain can quickly turn underground streams into torrents, trapping and sometimes drowning imprudent underground explorers. In most tourist areas, forecasts for inshore waters and local ranges of hills and mountains are available; they are an essential planning component.

Quieter leisure activities such as gardening, golf, athletics and all outdoor spectator sports are affected by the weather, and forecasts have a role to play. For several years special forecasting services have been provided for important sporting occasions. At Wimbledon, for example, the use of rainfall radar to accurately predict showers has enabled the grass tennis courts to be covered in good time ahead of showers. Many other outdoor sports such as baseball, football, racing and athletics, and especially cricket rely on a pitch or track being playable. When bad weather has already left its mark, the decision whether or not to postpone an event is often made in consultation with a weather forecaster. An early decision can save many thousands of supporters a fruitless journey.

All-weather tracks for athletics and horse racing, and indoor stadiums for other sports, enable events to take place regardless of prevailing weather, adding to their attraction. A striking example was the construction in 1965 of the Astrodome in Houston, Texas. In its first year of operation this huge, comfortable, air-conditioned arena drew more than three times the number of spectators to the regular baseball programme compared with the previous year. Underground heating prevents frost on most major rugby football and soccer grounds.

The economic benefits from weather forecasts in the fields of sport and leisure are unquantifiable. Amongst the more adventurous, many accidents are undoubtedly forestalled and lives saved. The rest of us benefit by being able to plan our activities to gain maximum enjoyment and suffer the least possible inconvenience from the weather. In 1994 the provision of 'free' public service weather forecasts in the United Kingdom cost less than 70 pence for each UK resident. The reader must judge whether or not this was good value for money.

——— # Health and social aspects ———

Our housing, dress and way of life all relate to the weather and climate of our homeland. We counter variations, day to day and season by season, using well-tried strategies – hats, coats, wellington boots, umbrellas, central heating, iced drinks, fans, shorts and swimming pools to name but a few. However, there are ways in which weather affects us of which we may be unaware. Comfort is a powerful influence on behaviour. Hot, sultry days and nights, can lead to loss of sleep and to physical and psychological stress. Occasional business visitors to tropical regions are well advised to insist on air-conditioned hotel accommodation, at least until they become acclimatised.

It has been found that heat and high humidity often accompany riots and civil disorder, especially in temperate climates. Furthermore, opportunistic crime increases in hot weather, primarily because windows and doors are much more likely to be open or unsecured to encourage ventilation.

High humidity combined with heat is particularly debilitating because our cooling mechanism relies on the evaporation of perspiration, which cannot proceed efficiently if the air is already nearly saturated. Where these conditions regularly occur air conditioning is essential, especially in hospitals where it will hasten the recovery of patients, in offices and factories where it will increase the output of workers, and in shops and restaurants where it will enhance the attraction of businesses to their customers. Not all hot climates are uncomfortable. Many desert regions have generally low humidity, with cool nights making a pleasant counterweight to hot dusty days.

Weather and climate play their part in many diseases. Some, like malaria, depend on a carrier such as mosquitoes thriving only in certain climates and terrain. Asthma and other respiratory disorders are often partly allergic reactions, and increase when pollens are concentrated in the air. This usually happens when pressure is high and winds light during the late spring and summer, when plant growth is strong and pollens concentrated in the lowest layers of the atmosphere. Breathing difficulties may also be triggered by an abrupt change to cold weather, or by fog and pollution which exacerbate bronchitis. The reduction in smoky fogs, or smogs, in the United Kingdom since clean air legislation was enacted in the 1950s has been dramatic, but pollution remains a serious problem (see Chapter 11). Viral diseases such as influenza and the common cold primarily occur in the winter half-year. Their frequency increases when a warm spell follows cold weather. Although vitamin D from sunshine is essential, skin cancer is highly correlated with excessive ultraviolet radiation from the sun. Decreasing stratospheric ozone, discussed in Chapter 11, is a factor in this.

Mortality directly due to cold weather, or **hypothermia**, occurs from time to time when outdoor workers or, more usually, winter holidaymakers are ill prepared. Hypothermia is sometimes called death from exposure. Anything that increases physical stress will exacerbate coronary disease, and cold winters increase deaths over much of

Europe and North America, especially amongst the elderly. In warm climates, including the southern United States, the death rate reaches a maximum in the hot summer months. **Heatstroke** occurs when exertion in high temperatures induces excessive perspiration and depletes essential body salts. Further sweating becomes impossible and the body temperature increases further, leading to coma and possibly death in the absence of medical aid.

9

WEATHER LORE

There will hopefully always be a place for tradition, custom and folklore in our lives, but it would be less than honest to claim that they always sit happily with science. Weather lore is almost as old as the human race and extensive, but much is of doubtful validity. It is always a poor substitute for the intelligent appraisal of up-to-date information. Where rules and sayings apply to weather in the longer term, say beyond the next few days, then experience and logic suggests they are worthless and indeed sometimes contradictory. Nevertheless, many sayings relate features of recent or present weather to the following harvest, and thus the future well being or otherwise of the population. This would clearly make a great deal of sense in times gone by, even if no longer valid today.

The strength and the weakness of weather lore is that it is based on human experience, and for each of us that is only a short thread in the tapestry of time. Our ancestors were crucially concerned with the impact of weather on their welfare; often it was literally a matter of life and death. Unfortunately, fluctuations in weather and climate tend to occur in epochs of all timescales. Thus, for example, if a similar sequence of weather events has preceded most poor harvests over a few decades, then the apparent relationship may enter folklore even though there is no causal link. The 'rule' is subsequently passed down through generations, being reinforced every so often when it happens to recur, but with contradictory examples conveniently forgotten (another facet of human nature). Attempts in recent times to use similar comparison methods to produce monthly and seasonal forecasts have also been misguided by epochs (see long-range forecasting in Chapter 7).

Weather lore is often not predictive of subsequent weather, even though it may appear so. Take, for example, the saying *All the months in the year, curse a fair Februeer*. This is a shrewd assessment of likely horticultural problems following a mild February, almost regardless of subsequent weather. Early warmth encourages tender crop growth which is susceptible not only to spring frosts but also to the infestations of insects which are likely to be earlier than usual. Before the days of chemical sprays, mild Februarys must often have led to worse than usual harvests later on, and the availability of food next winter.

Many sayings translate a reaction in nature to current or past weather into a forecast of what is to come. For example, plants that close their petals as humidity of the air becomes high, have been regarded as a predictor of rain since the time of Pliny. This has an element of truth, because approaching warm fronts often bring moist and cloudy winds well before rain starts to fall. On the other hand, the petals also close in the early morning mists of fine weather, and will rarely close before rain showers that often fall through quite dry low-level air.

Dubious weather lore sometimes appears to persist because of a desire to believe in magic or the supernatural; this is also seen in the widespread acceptance of astrology. For example, a bountiful harvest of wild berries is still popularly believed to foreshadow a hard winter, despite overwhelming evidence to the contrary. They are plentiful because the preceding spring and summer have been mild and moist; it has no bearing whatsoever on the following seasons.

There are a large number of supposed relationships between birds, animals and insects and the forthcoming weather. Invariably these reflect reactions to current weather conditions and it is from this that any predictive element arises. Even then any value is strictly short term. For example, the idea that birds (rooks are most often quoted) build nests higher before summers of light winds, and lower before windy ones is incorrect, though no doubt windiness during nest building will have an effect.

Astronomical configurations have neither proven value nor logical reason for having any value in weather forecasting. It is true that there are tides in the atmosphere as well as the oceans, but any relationship with weather is negligible (see Chapter 7).

The following is a much-abridged selection of weather lore together with possible explanations. The sayings comprise only the best known, plus a few others for which there is a tenuous scientific basis. They do not stray too far from modern English; early written versions of many sayings are indecipherable other than by historians!

When March comes in like a lion it goes out like a lamb, when it comes in like a lamb it goes out like a lion.
March is at the end of the long, northern hemisphere winter when the sun is moving northwards over the equator, and the temperature contrast between high and low latitudes is at its greatest. This means that the strength of the westerly circulation is also near its maximum, and vigorous depressions bring plenty of wet and windy spells across the British Isles and Western Europe in most years. The saying requires flexibility as to what precise periods are meant by 'coming in' and 'going out', but it does reflect the changeable nature of spring weather with a high probability of different types of weather early and late in the month.

Rain before seven, fine before eleven
Four hours of continuous rain occur occasionally in temperate latitudes when a weather front becomes slow moving, perhaps with waves (see Chapter 5), but periods of rain are usually shorter. 'Seven' and 'eleven' were perhaps chosen because other four hour periods do not scan well, or maybe it encouraged our rural ancestors to prepare to get out to work even when the early morning weather looked distinctly unpromising.

Long foretold, long last; short notice, soon past.
This is true insofar as the slow advance of a depression with falling pressure and thickening cloud brings bad weather often lasting a day or more, whereas rapid clouding over is more likely to be the precursor of a single shower or a smaller fast-moving depression.

Red sky in the morning, shepherd's warning; red sky at night, shepherd's delight
Near sunset and sunrise rays from the low sun travel a much longer path through the atmosphere than at other times, encountering many more dust particles. These scatter the longer wavelength red end of the light spectrum. When the sun is high in the sky, the scattering of shorter wavelength blue light by the much smaller air molecules predominates. A red sky at night suggests a clear sky for hundreds of miles beyond the western horizon, and no imminent frontal systems bringing rain. Red sky in the morning tells us little about what is

approaching from the west, but only that eastern skies are largely clear. However, when this corresponds to the zone of fine weather which often exists between two depressions, then it suggests it is moving away eastwards with more bad weather not far away to the west.

Ice in November to bear a duck, the rest of the winter'll be slush and muck

Winter in the United Kingdom rarely sets in before December, the coldest months are usually January and February. A very cold snap in November is unlikely to last, and will usually be followed by a milder spell with a thaw. November's weak sun will dry the ground only slowly if at all, and 'slush and muck' aptly describes conditions underfoot in the countryside when mild wet weather follows snow and ice.

Warm October, cold February

This, and all similar relationships between months may occur more often than chance expectation over some epochs and less over others. They have no long-term predictive value and such sayings are invalid.

Mackerel sky and mares' tails, make tall ships carry small sails

High cirrus clouds often form well ahead of depressions and their associated fronts. Mackerel skies and mares' tails describe forms of cirrocumulus and twisted sheaves of cirrus respectively implying strong high-level winds. In the days of sailing ships they will have been rightly viewed apprehensively as likely forerunners of stormy weather.

Evening red and morning grey, two sure signs of one fine day

'Evening red' is another way of saying 'red sky at night' (see above). 'Morning grey' probably refers to early morning mist, fog or shallow, low cloud (stratus) which often form on clear, near calm nights, but soon disperse after sunrise on a fine summer day. More extensive cloud especially with strong winds prevents the red sunset and overnight mist or fog, and of course a fine day is much less likely.

If clouds be bright, 'twill clear tonight. If clouds be dark, 'twill rain – d'ye hark.

Bright clouds suggest sun shining through gaps between and on to cumulus clouds, which result from the sun heating the ground when the atmosphere is unstable. These clouds often dissolve towards sunset to give a clear, cold night. Dark clouds are usually deep and more extensive. They do not usually vary much with local solar heating or the lack of it, and often bring rain.

A green Christmas makes a fat churchyard
A cold May gives full barns and empty churchyards
The first of these sayings may refer obliquely to an increased spread of infections, in years gone by, when a mild Christmas encouraged the more susceptible old and young folk out to worship and celebrate. The second may be explained by cold, wet springs encouraging sturdy if slow crop growth and discouraging early insect pests. These strong plants may have produced better harvests in those years, leaving more to store through the following winter and better nutrition for the populace.

Cast not a clout 'till May be out
'Till April's dead, change not a thread
A clout in this context is a piece of clothing. Both sayings date from when a single set of clothes would be worn throughout the winter. They simply mean that April and even May can be cold, so do not be misled by a warm spell.

When trout refuse bait or fly, there is ever a storm a nigh
The likely origin of this is that if fish are biting fishermen are oblivious of the weather; if they are not then fishermen blame it!

Turkeys perched on trees and refusing to descend indicate snow
It is more likely to indicate the proximity of Christmas, intelligent birds, and nothing to blame the problem on but the weather.

Cows lying down is a sure sign of rain
It is not even a sign of tiredness. Cows regularly lie down to chew the cud.

In by day. Out by night. (of wind)
A succinct description of the diurnal change between sea breezes blowing in from the sea during the day and land breezes out to sea overnight, explained in Chapter 5.

A bright circle round the sun denotes a storm and colder weather
Thin cirrostratus cloud often produces both solar and lunar halos (see Chapter 10). It precedes depressions which bring rain and strong winds, and eventually colder weather to the rear.

A northern air, brings weather fair
The north wind doth blow, and we shall have snow
Over the United Kingdom a north wind may precede the fine weather of a ridge of high pressure moving in from the west, or it may be associated with low pressure over the near continent or North Sea. In the first case, any showers soon become restricted to coasts and hilly areas in the north, but winter northerlies can bring many snow showers especially over northern and eastern regions. The sayings could perhaps be usefully combined as follows: 'if a north wind doth blow, then more we need know'!

Always a calm before a storm
Frequently, but not always, thunderstorms develop during hot sultry afternoons when there is little wind, but once formed they produce strong downdraughts and squally winds.

The sudden storm lasts not three hours
The sharper the blast, the sooner 'tis past
These both truly reflect the difference between the sudden heavy deluge and squally winds from a heavy shower or thunderstorm, and the generally steadier and often prolonged frontal rain associated with depressions.

When the clouds go up the hill, they'll send down water to turn a mill
This may stem from the increase of cloud first seen over hills as moist southerly winds pick up ahead of a depression. The mill in this case would be a water mill.

Oiled floors become damp before rain
This is one of many sayings that reflect the increased humidity in the air which often precedes rainy weather. Water condenses on to cold surfaces as air in contact is cooled below its dewpoint. Thus stone floors and walls in unheated buildings and with little or no covering will become damp. Pliny noted that 'if metal plates and dishes sweat it is a sign of bad weather'.

If the ash is out before the oak, you may expect a thorough soak.
If the oak is out before the ash, you'll hardly get a single splash.
This saying is contradicted by the following:

If the oak is out before the ash, 'twill be a summer of wet and splash,

But if the ash is before the oak, 'twill be a summer of fire and smoke.
What is the truth? In most years one or other applies!

If it rains on St Swithin's Day – 15 July – then we shall have rain for forty days

This is nonsense as even a glance over a few years' records shows. The legend has it that St Swithin died in 862 and was initially buried outside in accordance with his wishes. About a century later, on 15 July, he was reinterred inside the church. It is likely to have been wet on that day and for most of the forty days following, with superstitious minds quick to connect this with his displeasure. Not only is it surprising that the legend has persisted, but also that similar ridiculous sayings exist in other European nations: in France St Medard on 8 June; in Belgium St Godelieve on 27 July; and in Germany the day of the Seven Sleepers on 27 June.

A piece of kelp or seaweed hung up will become damp previous to rain

Any truth in this probably hinges on salt remaining on the surface of the weed. Salt is hygroscopic, which means it will absorb moisture when the air is humid. This may mean the chance of rain is slightly higher, see 'Oiled floors become damp before rain' above. Other observations in weather lore rely on the effect of increased (or decreased) humidity. Sailors noted that ropes tend to be harder to release ahead of rain (they shrank). Musical stringed instruments sound purer as tension increased due to shrinking. Rush matting was found to shrink in dry and expand in wet weather. The expansion and contraction of hair and gut have long been used by meteorologists to measure humidity, see Chapter 6.

10
WEATHER PHENOMENA

Many machines work quietly and unobtrusively throughout their useful life; others produce sparks, bangs and flashes which are rarely good news. Our weather machine is one of these latter! In previous chapters the main weather components have been assembled and what remain are predominantly local effects, but important for all that. We all marvel at the beauty of a rainbow, but even the bravest cannot help flinching at sudden cracks of thunder and flashes of lightning. Here these phenomena and others are described and explained.

Electrical phenomena

Thunderstorms

Thunderstorms are simply showers accompanied by thunder and lightning. They only occur with deep convective clouds called cumulonimbus, which often extend from within 1 km (0.6 miles) of the ground to well above 10 km (6 miles). Within these clouds there are great up and down currents, and a mix of water droplets, ice crystals and hailstones. These are of different sizes and weights, and are consequently lifted by, or fall through, updraughts in the cloud at different speeds. When small ice crystals collide and rebound from hail then they remove positive electrical charge. The heavier, now negatively charged, hailstones fall relative to the much smaller positively charged crystals which are carried up in the cloud. This results in what is known as **charge separation**, a massive difference in electrical potential or voltage between different parts of the cloud, and different parts of adjacent clouds and with the ground below. A deep and vigorous cloud with several cells with strong up and down currents will produce most frequent thunder and lightning.

Lightning and thunder

Lightning results from a gigantic electrical discharge, either between a cloud and the ground, or between points inside the same cloud or between adjacent clouds. It is exactly analogous to the flash which occurs when opposite terminals of a battery are almost connected, and a small spark or electrical discharge jumps the gap. This spark and a lightning flash mark the path heated almost instantaneously by the electrical discharge, to such a degree that it radiates visible light. The crackle of a small electrical discharge, and the loud thumping, banging and rumbling of thunder both result from the rapid expansion of air along the path of the discharge, generating sound.

The difference between sparks from a battery and lightning is one of scale. A car battery is usually 12 volts, while a typical lightning flash involves a difference in electrical potential of from 100 million to 1 billion volts. The highest electric gradients measured inside thunderstorms by rockets are a colossal 100 000 volts per metre.

Early lightning strokes in the lifetime of a thunderstorm are usually within or between clouds, often visible from the ground only at night as blurred areas of light, followed by muffled thunder. This is commonly called **sheet lightning**. A little later, often when or just before heavy rain or hail reaches the ground, the first lightning strokes occur between cloud and ground. This is when negatively charged melting hail falls close enough to the usually positively charged ground for electrical breakdown. These strokes are distinct jagged lines, where the lightning takes the line of least resistance through all the slight variations of temperature and humidity through the air. Studies show that often the first stroke from cloud to ground is followed almost instantaneously by a brighter return stroke in the opposite direction. This is probably due to ionisation (effectively loss of insulation) along the path of the first pulse, which immediately encourages a stronger discharge. Sometimes the flash takes several paths to ground from a main channel, this is called **forked lightning**.

At close range thunder is a startling loud crack rather than a rumble, as the sound is less distorted over a shorter path. Light from lightning travels at approximately 300 million m/s and may be regarded as instantaneous, but thunder travels at the much slower speed of sound, around 333 m/s. It follows that if the delay between a flash and the bang is timed in seconds, the result divided by three gives a good idea of the distance of the storm from the observer in kilometres, or if divided by two gives the distance approximately in miles. Sometimes, however, it is not easy to relate lightning strokes with their particular claps of thunder.

The huge electrical discharges of lightning interfere with radio reception, and in doing so, provide a means of locating storms thousands of kilometres away. Low-frequency radio pulses are induced by lightning strokes, and by accurately measuring the different arrival times of these pulses at receiving stations, the position of the parent storm can be accurately fixed. This technique is called **Arrival Time Difference** or **ATD sferics**. It enables thunderstorm activity to be tracked in real time over areas where few observations are available, especially the oceans.

Any electrical current or discharge takes the path of least resistance, and lightning is no exception, striking wherever the path to earth is easiest. This may be through the fabric of a building, in which case intense heating along its path can cause considerable structural damage or even fire. For this reason high buildings are always equipped with lightning conductors from their highest point to earth. These are thick copper rods or other good electrical conductors along which the lightning energy travels harmlessly to the earth. Even domestic houses on hill tops or isolated in flat surroundings are at risk, and precautionary external earthing of aerials or installation of special conductors is advisable.

The chances of a person being struck by lightning are small. In a thunderstorm one of the safest places is inside a motor vehicle, because any electric charge to earth will pass through the outside of the metal body and not affect passengers. This is known as the 'Faraday cage' effect after its discoverer. In built-up areas between or inside houses, shops or offices the risk is negligible. On the other hand, outdoors in a thunderstorm it is wise to have regard to the possibility of lightning strikes. Never choose to shelter under the largest of a group of trees, nor an isolated tree. Do not admire a storm from the summit of a hill nor in an open field or beach. If caught in the open certainly do not point a metal-framed umbrella invitingly at the sky, even if it means getting wet!

Ball lightning

This phenomenon is a rarely observed form of electrical charge concentrated in a dazzling sphere most commonly reported as being some 20 cm (8 inches) across. It occurs in thunderstorms, may persist for several minutes, and usually dissipates harmlessly. This phenomenon has not been satisfactorily explained. Some reports are almost certainly psychological in origin, resulting from a combination of the after image of earlier intense lightning flashes with anxiety. Others may result from electric charge becoming concentrated on a dust particle in a form of St Elmo's Fire (see below), or subtle chemical reactions on gases subject to extremely intense electric fields which produce a bright plasma.

St Elmo's Fire

This is a glowing electrical discharge from a metallic object which has been subject to a strong build-up of static electricity. It is most often seen on aircraft where it results from the impact of ice crystals in cloud, or from traversing areas of cloud (usually cumulonimbus) with widely different electrical potential. It appears as a blue/green glow and may persist for several minutes, but is rarely damaging beyond interrupting radio communications. Sometimes it is called a **corona**.

———— Optical phenomena ————

Rainbow

This most beautiful sight is the result of sunlight falling directly on to raindrops, so it is almost invariably seen on days of sunshine and showers. Every ray of sunlight falling on a drop is bent or refracted as it passes inside into the water. Some light is then reflected at the opposite face inside the drop and returns to be refracted again as it exits from the front (see Figure 10.1 (a)).

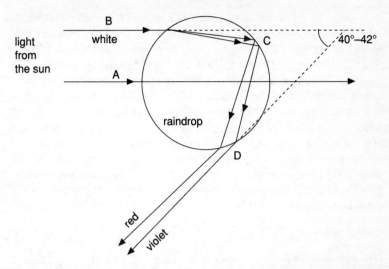

Figure 10.1 (a) Rays of light A and B from the sun encounter a spherical rain-drop. A, heading straight for the centre, is not deflected and passes straight through. B is refracted at the air/water boundary and part reflected inside the rain-drop at C, to eventually exit at D. Where it enters and leaves the drop, the light is split into component colours, because the different wavelengths are refracted through slightly different angles. This is similar to light split by a glass prism.

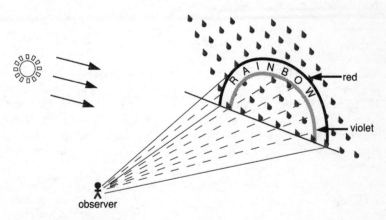

Figure 10.1 (b) Rainbow.

White light is a combination of seven colours each with a slightly different wavelength. A consequence of this is that each colour is bent differently on entering and leaving the drop. What enters as white light, exits as the complete visible spectrum of colours: red, orange, yellow, green, blue, indigo and violet, from long to short waves. To an observer viewing millions of raindrops in the distance, the whole spectrum appears side by side as concentric part circles (see Figure 10.1 (b)). The sky is brighter within the rainbow than without because light is concentrated by the reflection inside the drops.

A rainbow can only be seen if the observer is within 42 degrees of the line joining sunlight and the raindrops, because all the reflected and refracted rays return within this cone. It follows that if the sun is more than 42 degrees above the horizon no rainbow can ever be seen from the ground, consequently rainbows are much less common in the tropics (see Figure 10.1 (a)).

Some light is reflected twice inside each raindrop and emerges at a larger angle forming a secondary rainbow outside the primary rainbow with colours reversed and much fainter.

Since the spreading out of the light depends partly on the length of the path within the raindrops, the character of a rainbow depends on drop size. Drops greater than 1 mm (0.04 inches) produce brilliant rainbows, but small drops of less than about 0.05 mm (0.002 inches) are only faintly coloured at the edges. These are sometimes called **fogbows** because fog droplets are around that size.

Glory (or brockenspectre)

When the sun projects the shadow of an observer on to a bank of fog or cloud, perhaps from the top of a building or from an aircraft, then the shadow is often surrounded by coloured rings similar to those of a corona. This is caused by slight bending or diffraction of light at the shadow edge, the extent varying with colour across the spectrum. In this case each fog droplet acts as a tiny curved mirror reflecting the coloured halo effect back to the observer.

Mirage

The velocity of light through air varies slightly with temperature. Therefore, if the temperature changes considerably with height, light rays are bent or refracted towards or away from the ground. That is why over a hot sand or tarmac surface an observer often sees what appears to be a ghostly lake of water shimmering in the heat; it is a refracted image of the sky. This is called an inferior mirage because the image of the sky is below the real sky. Superior mirages are less common, but sometimes, especially above snow-covered surfaces, mountains will be visible when normally they would be over the horizon.

Solar and lunar haloes

Ice crystals form in different shapes such as needles, prisms, columns or plates at different temperatures. In slowly ascending air at high altitudes often a large area of thin cirrostratus cloud forms. The sun or moon viewed through such a cloud is often seen to have a bright circle or halo around it some distance away. Sometimes, especially at night, the cloud is so thin that this halo is the first indication of cloud's presence. This is important because the halo is a good predictor of rain ahead of a warm front (see Chapter 5).

The angle between the halo, the observer and the sun (or moon) is usually 22 degrees. This is because that is the minimum angle of refraction of light through an ice prism with faces inclined at 60 degrees, and high cirrostratus is commonly formed of uniform ice crystals shaped as hexagonal prisms. Haloes of 46 degrees do occur, but they require crystals with faces at right angles, which are comparatively rare.

Mock sun (Parhelion)

Like the solar halo this results from the refraction of sunlight by ice crystals, but in this case the ice crystals take the form of vertical needles, each of hexagonal cross section. The mock sun usually appears as two images of the sun, one on each side, often tinged red on the edge nearest to the sun. The angular separation is 22 degrees when the sun is low in the sky, but increases with higher elevation. **Mock moons** (Paraselene) also occur, but because light is much weaker they are rare.

Green flash

On rare occasions the last glimpse of the setting sun or the first glimpse at sunrise is brilliant green, perhaps accompanied by a green flash or spike shooting above the horizon for a few seconds. This is partly explained by the greater refraction of the green/blue/indigo/violet end of the spectrum enabling it to linger longest over the horizon. Since most reports are from tropical seas, it seems that high, low-level humidity and a strong inversion can sometimes completely, briefly, trap the blue end of the light spectrum. This is possible only at a low angle of incidence, such as at sunset, and is very similar to radar ducting discussed in Chapter 6.

Wind phenomena

Tornadoes and severe storms

Tornadoes consist of a rotating vertical cylinder of extremely strong winds, often exceeding 100 kn, varying from a few metres to exceptionally 100 m (330 ft) or more across. They are exceedingly destructive and dangerous because of the energy of the powerful winds and the debris thrown about. Also in the central core of a tornado pressure is greatly reduced and buildings crossed may have windows and doors sucked out, even if otherwise strong enough to withstand the blast. Tornadoes are produced in particularly intense thunderstorms with special characteristics, which fortunately in most parts of the globe are rare, they are called **severe storms**, or supercells.

The most common thunderstorm described earlier obtains its energy from below, where the sun or a warm sea heats low-level air in a particularly unstable or showery air mass, and generates deep cumulonimbus clouds. Normally each such storm lasts only an hour or so before its energy is spent. A severe storm is more long lived, usually deeper and more intense.

Sometimes a situation evolves where the air in the lowest kilometre or so is warm and humid, but is capped by an inversion or stable layer so that it cannot rise. In addition, winds higher up may be bringing much colder dry air across. The situation becomes like a balloon being blown up – sooner or later something has to give. Eventually, the inversion breaks down in a small area and within minutes a huge cumulonimbus cloud breaks through, soaring upwards through the troposphere. Dry winds aloft are swept aside by the updraught, but as soon as rain and hail start to fall through them from the cloud above, evaporation causes rapid further cooling. This generates strong down-currents which undercut the warm moist layer near the surface, scooping them up to reinforce the storm. Provided there is a continual supply of warm, moist air below and drier, cool air aloft both flowing towards the storm from different directions, then a severe storm composed of these so-called **supercells** can persist for many hours and travel hundreds of kilometres. Because the circulation within and around the storm continually regenerates it, it is sometimes called a **self-propagating storm** (see Figure 10.2).

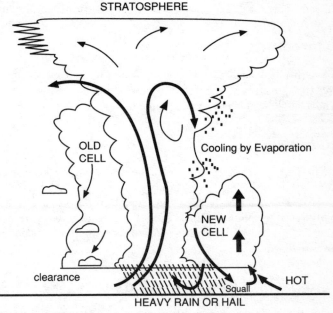

Figure 10.2 Schematic diagram of a severe thunderstorm.

Because low level air is trapped below a stable layer, the warm, moist air feeding the updraught into the storm is drawn from a wide area. This concentrates spin, in just the same way as water spins down a plug hole. In the early stages these centres of spin are weak, and may appear below the cloud as rotating, narrow, elongated 'V' shaped appendages called **funnel clouds**. Later, though, if favourable conditions persist they intensify and reach the ground as highly damaging tornadoes.

Tornadoes occur in many regions of the world but are rare near the equator. The most severe tornadoes affect parts of the southern United States where hot and humid low-level air flowing northwards from the Gulf of Mexico, over the Mississippi plains, encounters cold, dry, higher level north-westerly winds from the Rockies, often ahead of a cold front. Severe thunderstorms with intense tornadoes may then break out over a wide area, each producing a swathe of destruction up to about 1 km wide (0.6 miles wide) and sometimes travelling for 100 km (60 miles) or more. In the British Isles damaging tornadoes are reported a few times in most years, but are much more localised, short lived and weaker than their American counterparts.

Waterspouts

These are tornadoes but over water. They appear like a cylinder of water that seems to merge with the cloud above, and where it reaches the surface the water is very rough and disturbed.

Dust devils

Over deserts and dusty plains in the heat of the summer, convection plumes occur without cloud forming at all, because the air is so dry. One of these occasionally draws in enough low-level spin to form a rapidly rotating column of dust. This is a dust devil, whirlwind or, in Australia, a **willy-willy**. They are usually smaller and less damaging than tornadoes.

Anabatic and katabatic winds

Low-level winds are mostly dictated by the pattern of atmospheric pressure as outlined in Chapter 4. However, sometimes surface heating or cooling together with features of the local terrain have an overwhelming influence.

When air in contact with a sloping hill or mountain cools overnight, then it gradually drains down into the valleys as a cold current – this is a **katabatic wind** (see Figure 10.3 (a)). In most areas they are only light, and responsible for **frost hollows**, those valleys or dips in the ground where the minimum temperature on radiation nights is invariably lower than elsewhere, and fog most readily forms. In more

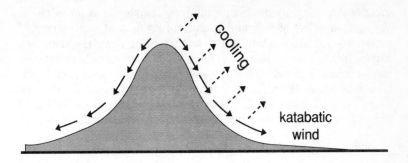

Figure 10.3 (a) Katabatic wind.

rugged terrain leading down to narrow fjords, and especially over the long snow and ice-covered slopes of Arctic regions, katabatic winds can exceed gale force. **Anabatic winds** blow up slopes heated by the sun usually following a cold night (see Figure 10.3 (b)). In that sense they are the opposite of katabatic winds, but are less common. They rarely exceed 5 kn because the heated air usually rises and breaks away from the surface.

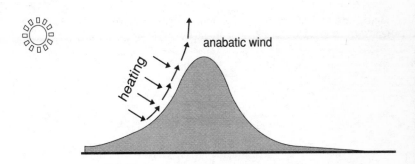

Figure 10.3 (b) Anabatic wind.

Föhn winds

When winds blow across a mountain range lee waves often form with or without wave clouds as discussed in Chapter 6. These oscillations occasionally bring dry air from well above the mountain tops right to the surface on lee slopes, and because of its descent and (adiabatic) compression it appears as a warm wind blowing down the slope (see Figure 10.3 (c)). Sometimes moist air lifted over the mountains deposits rain or drizzle on windward slopes and, being drier, warms more rapidly as it descends to the lee (see Figure 10.3 (d)). (It cools slowly when rising as cloud at the SALR, but warms more rapidly at the DALR on descent, see Chapter 3). Such winds are especially common where

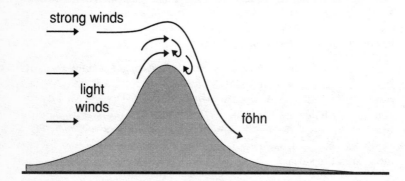

Figure 10.3 (c) Föhn wind resulting from lee waves bringing middle-level winds right down to ground level, warming by compression as they do so.

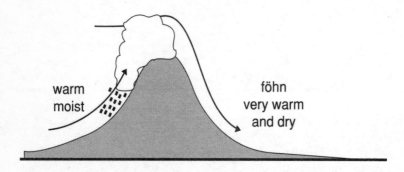

Figure 10.3 (d) Föhn wind resulting from a warm moist airstream losing rain and cloud as it is lifted up windward slopes, and consequently warming much more as it descends to the lee.

ranges of hills or mountains lie across the prevailing wind direction, for example the Rockies and Andes.

Local winds

The shape of topography and distribution of land and sea in many parts of the world combine with föhn and katabatic effects to give marked local winds in many places. Over the years these have been given special names. A few are given here:

bora   a cold, strong north-easterly through the Adriatic Sea: it is a combination of katabatic winds from mountains to the east and funnelling effects

buran   a cold strong north-easterly wind over central Asia, most prominent in winter when it occurs to the south-east of the dominant anticyclone over the continent

chinook   a warm, dry, westerly föhn wind to the east of the Rocky Mountains of North America: it is especially noted for the dramatic rise in temperature it can bring, an increase of more than 20 ° C in 20 minutes is not unknown

etesian   summer winds that blow southwards through the Aegean Sea: they are generated by the low pressure trough over the eastern Mediterranean in the lee of the Turkish mountains, together with funnelling effects

khamsin   a dry, dusty southerly wind over Egypt and the Red Sea induced ahead of depressions moving east through the Mediterranean Sea: an extension in the Sudan is called the **haboob**

leste (North Africa) and **leveche** (Spain)   hot, dry and dusty southerlies blowing from the Sahara desert ahead of depressions

levanter   a humid easterly wind that blows east through the Mediterranean on the northern flank of the north African summer heat low. It causes dangerous eddies near the Straits of Gibraltar

mistral   a cold, dry, northerly wind that blows across the Mediterranean coasts of France and north-west Italy, strongest where it funnels along the Rhône valley: partly katabatic from the Alps and Central Massif, encouraged by low pressure south of the Alps

pampero   a squally south-westerly bringing cold weather across Argentina and Uruguay, associated with cold fronts

sirocco   similar to the khamsin, but further west over the North African coast and Mediterranean: although originally dry, it moistens over the sea and brings low cloud into Italy and France

11

WEATHER AND THE ENVIRONMENT

Air pollution

The normal composition of the mixture of gases we call 'air' is given in Table 2.1, but unfortunately the air we breathe usually contains additional gases and particles, occasionally reaching concentrations which present a hazard to our well-being. This is what is meant by **air pollution** or **atmospheric pollution**. It does not always threaten our health directly but sometimes through our diet, affecting crops we grow and farm animals. Not all atmospheric pollution is manufactured; natural pollution from volcanic eruptions in the form of noxious gases and ash has occurred throughout geological time sometimes in vast quantities. Unfortunately, in recent years large-scale industrial processes, deforestation and the increased use of fossil fuel (gas, coal and all fuels derived from crude oil) have dwarfed natural production.

Air pollution does not just lower the quality of life, but shortens it. London was plagued by sooty fogs or **smogs** from the sixteenth century, when coal burning became widespread in what was already a high-density population. Episodes reached a peak in the notorious so-called 'pea-souper' of autumn 1952, when smoke mixed with fog was so dense pedestrians could not see their feet! Inevitably the city ground to a complete halt, and the smog caused 4000 deaths. Clean Air Acts, permitting only the use of smokeless fuel, have improved the situation immensely since that time, but newly industrialised nations face similar problems today. The benefits from decreased pollution accrue to all, but have an economic price. The political imperative to raise standards of living in the developing world understandably takes precedence over what is seen as a peripheral issue, at least until it becomes too serious to be ignored.

Pollution in the more prosperous economies is no longer focused on particulates, that is to say microscopic particles of soot which remain suspended in the air. Noxious, mostly invisible gases have recently taken centre stage, with the motor vehicle and its internal combustion engine presenting the greatest menace. These gases, in suitable meteorological conditions, become sufficiently concentrated to cause severe breathing difficulties to those with normally minor respiratory problems. Longer term effects on every one of us are as yet unknown, but they are not likely to be beneficial.

Main polluting gases, their source and composition

(C = carbon, F = fluorine, H = hydrogen, N = nitrogen, O = oxygen, S = sulphur)

Benzene C_6H_6

A hydrocarbon constituent of oil and gas, released both in the exhaust of badly tuned engines and by evaporation from petrol and aviation fuel. Implicated in some cancers.

Carbon dioxide CO_2

Not strictly a pollutant because it is a natural component of air. However, it is increasing because of the large-scale combustion of fossil fuels, the effect of which may be leading to global warming.

Carbon monoxide CO

Colourless, odourless and poisonous, carbon monoxide is a product of combustion. It poisons by preventing red cells in the blood taking up oxygen.

Chlorofluorocarbons CFCs

This class of manufactured compounds are cheap to produce, non-toxic and inert, and have until recently been widely used as aerosol propellants, refrigerator coolants, cleaning fluids and in the manufacture of plastic foam. They are strongly implicated in the destruction of stratospheric ozone.

Hydrogen sulphide H_2S

The main efflux from volcanoes, it is also produced in small amounts by catalysed engines. It smells unpleasant but is not poisonous.

Methane CH_4

Probably the commonest hydrocarbon in the atmosphere, methane is sometimes called marsh gas because it is produced by decay of organic matter, and is associated with boggy areas. It can build up in mines or buried refuse to give a dangerously explosive mixture.

Nitric oxide NO

Formed in small quantities by combustion and also from nitrogen dioxide (see below). In moist air it converts readily to dilute nitric acid.

Nitrogen dioxide NO_2

Forms from nitrogen and oxygen during combustion.

Ozone O_3

Although a natural and important constituent of the upper atmosphere, ozone at ground level is poisonous and also damaging to plants and natural materials such as rubber and textiles. Even small amounts irritate sensitive air passages and exacerbate allergies.

Sulphur dioxide SO_2

Produced by combustion from sulphur in fossil fuels and oxygen in the air, sulphur dioxide combines with water in the air or lungs to form dilute sulphuric acid (H_2SO_4). It is the prime constituent of acid rain.

– The meteorology of local pollution –

Pollution in the air normally presents a serious hazard only when it is concentrated. That requires weather conditions to be such as to limit dilution which would occur if there was mixing with unpolluted air. The first condition, therefore, is calm or light winds, because strong winds are effective at mixing. There is always a small amount of mixing due to the random motion of gas molecules (called molecular diffusion) but this is usually insignificant. The wind not only mixes air in the horizontal but also in the vertical, often through a depth of 1000 m (3000 ft) or more. The precise depth depends on the way the temperature changes with height, or the stability of this **boundary layer**.

Strong winds create turbulence in the form of a series of eddies, with air continually being forced up to be replaced by air from above. This mixing does not create a layer with the same temperature. As explained in Chapter 3, rising air cools and descending air warms at a rate of almost exactly 3 °C in 1000 m (the dry adiabatic lapse rate or DALR). A turbulent boundary layer, therefore, has this temperature lapse throughout the depth of mixing. In pollution studies it is regarded as **neutral stability** because air displaced in the vertical experiences temperature changes in line with its surroundings, and neither gains nor loses buoyancy.

In light winds the pre-existing temperature change with height, dictated mostly by incoming and outgoing radiation, remains almost unaltered. When the temperature falls with height at less than the DALR the boundary layer is said to be **stable**. This is because air displaced upwards cools faster than its surrounding environment, becomes heavier relatively, and will sink back to its original level. If the temperature falls with height at more than the DALR, then air displaced from near the surface remains warmer than its surroundings and will continue to rise, until constrained higher up. This is an **unstable** boundary layer.

The worst pollution events occur in stable conditions when the temperature rises with height, that is to say there is a temperature inversion. This often occurs on radiation nights with no cloud and little wind, when the ground temperature falls rapidly cooling the air next to it. With little mixing, smoke and other pollutants become trapped near the ground and concentrations increase. Amongst meteorologists an inversion is referred to colloquially as a 'lid', because it often caps cloud, fog or pollution. Tall chimneys are used in many industrial processes, so that waste gases and smoke will disperse above a shallow inversion, and hopefully stay above ground level. An inversion formed in this way may persist and intensify over several days in winter, when the weak sun cannot warm the ground sufficiently to break through the inversion and change the stability. Pollution events are most often associated with high pressure, not only because winds are usually light and skies often cloud-free in anticyclones, but also because they create inversions, as described in Chapter 5. In summer the sun is able to heat the ground by day sufficiently to destabilise the boundary layer releasing pollution to mix through a progressively deeper layer. Even so summer anticyclones are often hazy, especially where they persist for many days. Figure 11.1 shows typical temperature profiles associated with varying levels of pollution.

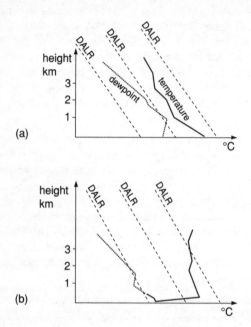

Figure 11.1

(a) Typical temperature and humidity on an unstable showery day. The lapse rate is greater than the DALR and air readily rises from the surface mixing pollution through a deep layer.

(b) A stable night-time profile of temperature and humidity, when an intense inversion near the ground traps pollution in the lowest few tens of metres.

Photochemical smogs

Where summer anticyclones are persistent, especially in the subtropics (see Chapter 2) a low-level inversion may exist for many weeks. This situation gives rise to **photochemical smogs** in landlocked cities. In bright sunlight a process called **photochemical dissociation** can remove an oxygen atom (O) from the two oxygen atoms of nitrogen dioxide (NO_2); this combines with an oxygen molecule (O_2) to form poisonous ozone (O_3), leaving a nitric oxide (NO) molecule. Hydrocarbons such as methane and benzene may each be similarly dissociated from one of their oxygen atoms. Vehicle exhaust gases contain the necessary cocktail of gases for this to occur, which explains the prevalence of such smogs in large cities in summer. However catalytic converters on vehicle exhausts remove almost all hydrocarbons and nitrogen oxides helping greatly to solve this problem.

If fog forms in polluted air, as is often the case in winter, each fog droplet soon becomes a dilute mixture of acids, presenting an even greater health hazard as well as eating away at the fabric of our civilisation.

Acid rain

Tall chimneys on factories and power stations may disperse their mixture of noxious gases high into the air to be wafted away in the winds, but they must go somewhere. With predominantly westerly winds in middle and high latitudes it is not surprising that areas east of large industrial areas are most seriously affected. As cloud and then rain forms in polluted air, sulphur dioxide dissolves to form dilute sulphuric acid, and oxides of nitrogen form nitric acid. The effects of such **acid rain** on vegetation and animal life are serious over a long period even at low concentrations.

Acid rain dissolves aluminium which is a natural but normally stable component in most soils. When it subsequently drains into rivers and lakes it proves a fatal mixture for fish and much aquatic life. During the last 25 years many rivers, lakes and lochs in rural areas of Canada, Norway and Great Britain have suffered severe depletion of fish stocks. Acid rain dissolves and washes away essential nutrients in the soil, such as calcium and magnesium, as well as liberating aluminium. Increased acidity damages vegetation, and many forests of North West Europe have tracts of dead and dying trees. Sulphur dioxide is also found to damage leaves and pine needles directly by blocking the stomata on leaves, thereby preventing life giving photosynthesis. Photosynthesis is the process whereby sunlight acting on chlorophyll in green leaves converts carbon dioxide from the air into carbon to grow the plant, and oxygen which is released into the atmosphere. Any depletion of forests reduces the absorption of atmospheric carbon dioxide, thereby contributing indirectly to global warming.

Global warming and the greenhouse effect

The composition of the atmosphere has changed considerably over geological time, as different life forms have evolved to comprise the present complex biosphere of animal and vegetable life. Such changes have been slow, and it is worrying that there is conclusive evidence that carbon dioxide (CO_2) in the atmosphere has increased considerably over only the past hundred years or so. Furthermore it continues to increase. This change is bound to have an effect on the radiation balance of the Earth, discussed in Chapter 4.

Shortwave radiation from the sun penetrates the atmosphere to warm the Earth's surface, which in turn warms the air in contact with it generating convection currents and hence heats the troposphere. The Earth and atmosphere radiate heat outwards in longwave bands much of which is reflected by clouds and certain gases, especially water vapour and CO_2. It is reasonable to suppose, therefore, that an increase in CO_2 will reduce outgoing radiation further, causing the temperature of the Earth to rise. This will increase outgoing radiation until equilibrium is again reached, but only after a net rise in global temperature.

This warming from increasing CO_2 has become known as the **greenhouse effect**. It is likened incorrectly to a greenhouse husbanding heat, whereas a greenhouse maintains higher air temperature than its surroundings by preventing air within from mixing with that outside. Experiments show that it has nothing to do with glass restricting outgoing radiation. In fact, pollutants such as chlorofluorocarbons, molecule for molecule have a greater effect in reducing outgoing radiation than CO_2, as do methane and nitrous oxide, but amounts of CO_2 in the atmosphere vastly exceed all of these together, making it by far the most important.

There is a huge store of dissolved CO_2 in the oceans, another in the form of carbon in forests and other plants, with yet another in fossil fuels, and until recently balance was maintained in the atmosphere. The increase of atmospheric CO_2 has come about through the massive increase in the use of fossil fuels and the burning of vast areas of forest, especially equatorial rainforest. The oceans will almost certainly absorb more CO_2 as its concentration in the air increases, and plant growth generally may benefit. However, while our profligate use of fossil fuels may be but a blip in geological time, there is no prospect that natural processes will balance the increase in atmospheric CO_2 in our lifetimes, nor even in those of our great grandchildren's children. What then is the effect likely to be?

If there is a small rise in average global temperature, much of the retained heat will be absorbed in raising the temperature of the oceans by a small amount. This minute increase will inevitably lead to more evaporation and increased cloudiness, but not everywhere. This will tend to reduce further warming by reflecting more solar radiation straight back into space. When the sun is high in the sky the sea reflects only about 5 per cent of incoming radiation, whereas most cloud reflects more than 50 per cent. Some scientists expect sea levels to rise both from melting ice and expansion of the warmer seas, but even this is open to doubt. First, the melting of sea ice can have no effect – it already displaces its own weight of water (this is Archimedes' principle). Second, increased evaporation from warmer currents travelling polewards may increase snowfall over Arctic regions, even if permanently frozen areas become less extensive.

The Earth and atmosphere form a robust system, not easily shaken off balance. What to us are major environmental emergencies – forest fires, earthquakes, Dutch elm disease and oil spills – are as a water drop on the back of an elephant. Expanding population and industrialisation may have a permanent impact, but meteorologically it is rarely possible to determine cause and effect. There have been past global variations of temperature, rainfall and atmospheric composition over all timescales, and it is likely that short-term changes due to the activities of humans will be small, though possibly significant. Even the measurement of global average temperature is subject to large uncertainties. Weather systems one year may steer warm air frequently over our limited sample of stations, with the inevitable gaps in our observation network being colder than usual. In those circumstances, a warmer than normal average is purely an accident of sampling. Furthermore, we know that there are epochs in meteorology; they lead to clusters of years biased in one way or another much more than would be expected from random or statistically normal distributions. The measurement problem is being solved by external observations from satellite and it now seems almost certain that the increase of 'greenhouse' gases, especially CO_2, over recent years has led to a small increase in average global temperature.

If measurement is difficult then prediction of climate change is doubly so. Attempts have been made by adapting the powerful numerical weather prediction (NWP) models (see Chapter 7) used in operational weather forecasting. Some of the fine-scale detail has been removed, and longer-term variables added such as effects of the interaction between oceans and atmosphere and variation of atmospheric gases, as far as current knowledge allows. Using these climate models, possible long-term variations of global weather over the next century or beyond can be projected using different assumptions, and the results compared. These models are greatly simplified version of the real world, and even if they produce plausible results when run with and without increased CO_2, it is not possible to say definitely that the results are correct. Perhaps the best we can hope for is to derive a balance of probabilities; at present that points to a likely global temperature increase of order 1 °C with every doubling of CO_2.

Ozone layer depletion

The **ozone layer** normally spans the globe high in the stratosphere, concentrated around 25 km (15 miles) where it lies above about nine-tenths of the atmosphere. It is important because it shields the Earth from damaging ultraviolet radiation from the sun. Ultraviolet radiation is dangerous to life, indeed it is used as a medical sterilising agent to kill microorganisms. The small amounts that pass through the ozone layer are responsible for sunburn, some skin cancers and many eye problems such as cataracts. It follows that any depletion of stratospheric ozone will adversely affect humans, animals and crops. Just such a decrease in ozone has been discovered over the past few years, and is of great concern.

The ozone layer forms because of chemical reactions in the high stratosphere. Solar ultraviolet radiation is so powerful at that height that ordinary oxygen (O_2) molecules are broken apart into two single oxygen atoms (O_1). These are reactive and readily combine with surrounding molecules amongst which are many unaltered oxygen (O_2) molecules, when the addition of an extra atom produces ozone (O_3). The ultraviolet energy absorbed in this process leads to a rise in temperature; which is why the temperature in the stratosphere increases with height, unlike in the troposphere below. Ozone creation, not surprisingly, is greatest in the tropics where solar radiation is strongest, but it is spread all round the globe by high level winds. The stratosphere is stable with little vertical mixing, hence the ozone layer does not mix down into the troposphere.

Ozone itself can be broken up by the absorption of ultraviolet energy of rather longer wavelengths, producing an oxygen molecule (O_2) and a free atom (O_1), but again the free oxygen atom will not exist long on its own. However, other reactions destroy ozone permanently. In particular oxides of nitrogen (NO and NO_2 molecules) convert ozone and free oxygen atoms (O_1) back into oxygen molecules (O_2) in a series of reactions. First, nitric oxide (NO) takes an oxygen atom (O_1) from ozone (O_3) to form nitrogen dioxide (NO_2), leaving an oxygen molecule (O_2). Second, nitrogen dioxide (NO_2) gives up one of its oxygen atoms (O_1) to combine with a free oxygen (O_1) atom forming an oxygen molecule (O_2) and leaving nitric oxide (NO), ready to repeat the first reaction. We thus have a circular series of chemical reactions which both break down existing ozone (O_3) and mop up free oxygen atoms (O_1) so preventing them from forming ozone.

Chlorine atoms are equally efficient at stripping the extra atom from ozone and so are fluorine compounds. What is more, just like the reactions outlined above with nitrogen oxides, chlorine molecules are recycled enabling each to break down many thousands of ozone molecules. The primary source of chlorine atoms in the stratosphere is from manufactured chlorofluorocarbons or CFCs. These are long lived and when released into the atmosphere gradually disperse, some inevitably reaching the heights of the ozone layer.

The generation and destruction of ozone in the stratosphere reaches a natural balance, which varies with latitude and season as solar radiation changes. The excess formed at low latitudes is transported polewards by winds in the stratosphere. In normal circumstances a blanket of ozone would span the Earth through the year, varying in density with the intensity and duration of daylight. CFCs have markedly altered things.

Regular measurements of ozone have been made by scientists of the British Antarctic Survey since the 1950s at Halley Bay in Antarctica using a **Dobson photometer**. This instrument analyses the total ozone above by measuring the intensity of its known effect on certain wavelengths in the spectrum of sunlight. Until 1982 there had been the expected reduction in ozone following the long dark months of the Antarctic winter, but in that year researchers first discovered a hole in the ozone layer. Since then ozone has always disappeared over a wide area in the Antarctic spring, recovering slowly as summer approaches. The hole has been verified by measurements from aircraft and satellite, and is getting bigger. In some years its area exceeds that of the United States. Recently evidence has been found of a similar depletion of ozone in the Arctic spring. There is little doubt that effects carry over into summer across a much wider area of both hemispheres, and that it will get worse for many years, even when (if) the use of chlorofluorocarbons has ceased totally.

The reason behind the sudden decrease of ozone in spring at high latitudes is complex and not fully understood. The extreme cold of Arctic and Antarctic winters extends to the stratosphere, where ozone formation with its consequent heat release ceases because there is no solar radiation. The low temperatures, down to -90 °C, cause extremely nebulous ice clouds to form, and these appear to react with chlorofluorocarbons to release free chlorine atoms. These readily combine with oxygen or hydrogen initially, but when sunshine returns in the spring these simple compounds are quickly broken down into the form in which they destroy ozone. This they do on a massive scale. Eventually, full summer sunshine creates ozone fast enough to outweigh destruction; the ice clouds are by then long gone and the chlorine atoms are gradually lost to more stable compounds.

It has been known for many years that gases released by volcanic eruptions give rise to a reduction of stratospheric ozone as they diffuse into the stratosphere. However, such changes have always been temporary. Current evidence suggests it is vital that the CFC pollutants responsible cease to be produced, and are never released into the atmosphere. Governments have already banned their use as aerosol propellants, and their use as cleaners and refrigerants is being phased out. Old refrigerators and air conditioners must have their CFC coolants removed carefully for chemical destruction, rather than being dumped and the fluid left to leak into the atmosphere. The long-term consequences of drastically reduced ozone are not known. It is expected that they will include many more skin cancers and eye defects, even though humans will learn to cover up in bright sunshine. The fate of animals and crops is even less certain.

12
CLIMATES OF THE WORLD

──────── **Weather and climate** ────────

Short-period variation of temperature, wind, humidity and rainfall comprise the **weather** at a location. Over much of the world one day is often quite different from the one before, and for many purposes a more general overall assessment of weather is required, summarising what has happened in the past and might be expected in the future. For example, anyone contemplating an overseas visit to unfamiliar territory wishes to know what conditions to expect, without having to wade through masses of daily weather records for previous years. They require a summary giving average values at the relevant time of year, and perhaps also extreme values of say, maximum and minimum temperatures, rainfall, sunshine and humidity. Such seasonal or monthly averages for the whole year describe the **climate**. This is by no means a substitute for a weather forecast, but gives an excellent indication of likely conditions the traveller must be equipped to face.

In some parts of the world temperature and rainfall vary considerably day by day, and between night and day, while in others they are almost constant. This variability is an important part of the climate, and so is the likely frequency of severe weather. A building designed for a city where temperatures rarely fall below 20 °C and often exceed 40° C needs to be different from another required to protect its occupants from severe gales in temperatures well below freezing. Fortunately, the main factors of the large-scale circulation discussed in early chapters, together with geography, have a large influence on climate. And this means that large areas of the globe lie within well-defined climatic zones, as we shall see.

The word 'climate' comes from the Greek *clima* meaning 'inclination of the sun'. It does depend mainly on latitude and hence the power of the sun, together with the general circulation of the atmosphere described in Chapter 2. However, the geography of each location in relation to its surroundings is also important. Here we are concerned only with the large scale or **macroclimate**, across countries and continents, not small local variations. For example, large towns stay warmer at nights than nearby rural districts, because buildings are reservoirs of heat. This so-called **heat island** effect is part of the local **mesoclimate**. On still smaller scales the fact that the shaded north side of a tree favours moss and hence particular insects is part of the **microclimate** relating to that particular tree or patch of woodland.

Effects of hills and mountains

Average temperatures invariably decrease with height, and the climate on mountains will always be colder than that in nearby valleys. There are, of course, occasions on which temperature readings show the reverse of this. For example, ahead of a warm front valleys may stay colder than hill tops for an hour or two, while often following cold radiation nights slopes facing the rising sun warm faster than valley bottoms. But these are exceptions, and climate is concerned with long-period averages. In the tropics it is possible to escape the hottest summer weather by spending the season in the mountains, while on relatively small islands such as Cyprus it is possible to ski on the mountains in the morning and bathe from hot beaches in the afternoon!

Windward slopes – usually those facing the west or south-west – invariably have greater average rainfall than those to the lee. The effect is marked even with modest ranges of hills or mountains such as in the British Isles, but where large ranges of mountains lie at right angles to prevailing westerly winds, such as the Rockies of North America and the Andes of South America, the results are dramatic. Moist winds forced to ascend in crossing the barrier cool, forming cloud and often rain which falls mostly on western slopes and the mountain tops. By contrast, east of the mountains the air is much drier and warms as it descends; consequently lee slopes and large areas beyond see little rain and higher daytime temperatures. This is the **rain shadow** effect.

Effects of continents and oceans

Land surfaces are poor conductors of heat; their temperature increases rapidly in sunshine and decreases equally quickly at night. The sea is different, with little change of surface temperature from day to night (see Chapter 4), except where it is exceptionally shallow.

The same applies to seasonal changes. Through autumn and early winter the sea is, in effect, a huge reservoir of heat, maintaining adjacent coastal areas much milder than regions well inland. In summer, on the other hand, it provides cooling sea breezes often keeping average temperatures near coasts markedly below those produced inland. Because of this there is generally a limited diurnal and annual range of temperature on small islands in the middle of oceans, but extremely large differences in the interior of continents at the same latitude. Between these two extremes the climate of every place on Earth is influenced to a greater or lesser extent by its position in relation to sea and continent. For example, winters in South East England are colder than those in north-west Scotland. This is because western Scotland is bathed by the warm water of the Gulf Stream and sheltered from cold continental winds from the east by the Grampian mountains; South East England is surrounded by land apart from cold and narrow stretches of the North Sea and English Channel. Even smaller islands have marked variations. The subtropical Canaries off north-west Africa have an ideal agricultural climate in the north and on their mountains where north-westerly winds bring showers, but a few kilometres away their southern parts are dry and hot, making them attractive to tourists from northern Europe, but little else.

——— Climate classification ———

The following simple climate classification is approximate, since there are marked differences within most climatic zones (see Figure 12.1). In addition, every year is different, and many boundaries cannot be allocated precisely; for example, the occasional long hot summers over north-west Europe are sometimes called Mediterranean-type summers, because they more nearly reflect that climate. Despite these qualifications, this framework gives an insight into the rich variety of cultures and lifestyles around the world, as they are influenced by the weather.

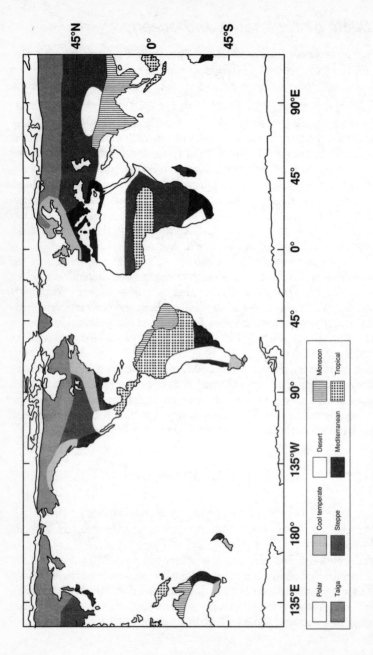

Figure 12.1 A simple climate classification across the world.

POLAR

Very cold winters, cool summers, dry throughout.

This is the harshest climate on Earth, and occurs within the Arctic and Antarctic circles. It is cold all the year round and extremely so in the long, dark winter months. The lowest temperatures are recorded in these areas, the record being –88 °C at Vostock, a Russian research station in the Antarctic. While high pressure is often maintained for periods of weeks, deep depressions bring strong winds, snow and blizzards at all times of year, but especially during the spring and autumn.

Little life survives the winters, and there is no permanent vegetation, but during the brief summer many types of small plants and millions of insects enjoy their brief life cycle especially in the Arctic, and migratory birds fly in to take advantage. Year-round natural inhabitants are mostly restricted to seals, sea lions, penguins in the south, and polar bears in the north. There is no farming, but fishing grounds yield high catches during the summer as the ice retreats.

TAIGA OR COLD FOREST

Cold dry winters, warm dry summers

This climate is peculiar to the northern hemisphere where the largest continents in the world extend to the Arctic. The huge expanse of Canada and even greater breadth of northern Russia, mean that most parts are far from the influence of mild winds off the sea. Consequently, the climate in these vast landlocked countries is one of extremes, characterised by short, often hot, summers but much longer, intensely cold winters. Rainfall amounts are generally small. In winter, pressure is usually high but occasional depressions bring snow from the north-west especially early and late in the season. In summer, there are long dry periods but also spells with showers and thunderstorms. The annual range of temperature is greatest in these regions: Verkhoyansk in Siberia has recorded –70 °C in winter and +37 °C in summer.

Vegetation consists almost entirely of conifers, which survive extreme cold and are not easily damaged by snow. ('taiga' is from the Russian word for 'forest'.) Animal life has been plentiful, particularly of large deer, brown bears and, in Russia, wolves; but in recent years many species have suffered from indiscriminate hunting.

STEPPE

Cold wet winters, hot dry summers

South of the cold forest belts in the interior of northern continents it is also dry for much of the year; winters are less extreme although still cold, but summers are long enough to give a productive growing season. This steppe climate is named after the vast plains or grasslands that cross Russia and the Ukraine, but applies also to the prairies of Canada and the United States, the semi-arid veldt of southern Africa, the downs of south-eastern Australia, parts of North Island, New Zealand and the pampas of southern Argentina. The southern hemisphere steppe climate is warmer and generally less extreme, because the plains are either at lower latitudes or nearer the sea. Over the northern steppe, rain falls in the summer months when cold fronts push southwards, triggering showers and thunderstorms; in winter depressions occasionally bring blizzards. The southern steppe climate sees little or no summer rain but winter depressions usually make up for that.

These can be productive agricultural areas, particularly for cereal crops over North America and the Ukraine. Unfortunately in some years they are subject to drought and consequent crop failure.

TEMPERATE (sometimes cool temperate)

Cool winters and warm summers, with rain in all seasons

Temperate regions are those experiencing the most *weather*; that is to say they lie in the latitudes of the changeable westerlies. Weather systems bring periods of rain throughout the year, which penetrate well into the interior of continents such as Europe which has no major mountain barrier. They include Northern Europe, British Columbia, Oregon and Washington State in North America, southern Chile, Tasmania and South Island, New Zealand. In addition, the north-eastern United States, much of eastern China and Japan are temperate, although these tend to be drier overall and colder in winter because they lie east of large continents.

In temperate zones latitude, and shelter by hills or mountains, probably have the most marked effect on mean temperature and rainfall. However, all these regions have a dormant season when the mean temperature is below 6 °C and plant growth ceases. Conditions are generally favourable for abundant animal and plant life, and it is probably no coincidence that approximately 80 per cent of the world's population live in temperate zones.

MEDITERRANEAN (sometimes warm temperate)

Mild wet winters, hot dry summers

Despite its name this climate is not peculiar to countries fringing the Mediterranean Sea. It also applies to parts of the United States including California, much of South Africa and southern Australia, northern parts of Chile and Argentina, and much of North Island New Zealand. While depressions bring winter rain, episodes are usually brief, with a notable feature being the prevalence of sunny days throughout the year. Summers are invariably hot with occasional thundery spells and little rain. Frost and snow are rare except over mountains.

This climate is ideal for many fruits, especially grapes, melons and citrus. Vegetables thrive, and often more than one crop a year is possible, especially where irrigation is available to supplement the meagre summer rain. Tourism is a major industry, especially in the northern hemisphere.

DESERT

Hot dry summers, warm dry winters

Desert regions occur chiefly in those zones of almost permanent high pressure and clear skies, between about 20 degrees and 35 degrees latitude, on the poleward side of the tropical Hadley cell. Nights can be cold, especially in winter because of the lack of cloud and low humidity; by day they are generally hot and arid. The most well-known deserts are the Sahara (which is nearly as large as the United States) and Kalahari of Africa, the Atacama of South America, the Great Western Desert of North America and the Australian Outback. The Gobi desert in central Asia is rather further north, the result of being completely landlocked for thousands of kilometres, and shielded to the south by the Himalayan mountains. The smaller deserts of Patagonia in Argentina, and in Nevada and Utah in the United

States also lie further from the tropics; they are entirely due to the rain shadow effect of mountains to the west. Even so, they see more rain than most other desert areas.

Vegetation is sparse, though rapidly triggered by the rare showery spells. These can also initiate the life cycle of dormant insects such as the avaricious locust, the eggs of which sometimes lie for years in the dry sand. Irrigation can turn desert into productive land, such as that near the river Nile as it flows through Egypt and the Sudan. Islands in desert latitudes enjoy occasional showery spells in the winter; which makes them ideal climates for agriculture as well as for holidaymakers. Their climate may be best described as Mediterranean, although even small islands such as the Canaries, mentioned earlier, may be desert in the south and Mediterranean in the north.

TROPICAL

Hot all year, wet summers, dry winters

In most climates there are marked variations between the seasons, but close to the equator this is not so. Every day the sun is nearly overhead so it is hot all the year round. The intertropical convergence zone or ITCZ (see Chapter 2) in its wanderings north and south of the equator brings copious rainfall, but often there are two or three dry months in the winter half-year when it is furthest away in the summer hemisphere.

Conditions are ideal for vegetation, and consequently the greatest rainforests of the world lie within 10 degrees latitude of the equator. The most well known are the Amazon rainforest of Brazil, the Congo basin of Central Africa, and the jungles of Malaysia, Indonesia, Burma and Vietnam. Animal, bird and insect life abound, not all of which is friendly to humans. But neither are humans friendly to them, because we are slowly destroying their environment, and in consequence perhaps upsetting the world's climatic equilibrium, discussed in Chapter 11.

MONSOON

Hot summers, warm winters, with a pronounced wet and a dry season

So far we have discussed climates that are governed mainly by the seasons of the year and the general circulation of the atmosphere.

Monsoon climates are different, because they are driven primarily by temperature differences between land and sea. They may be regarded as a large-scale manifestation of the sea breeze circulation described in Chapter 5. But whereas the sea breeze is a local and temporary phenomenon, a monsoon develops from land–sea temperature differences generated over many months and on a continental scale. The word 'Monsoon' derives from the Arabic *mausim* or season, and was originally applied by Arab fishermen to the seasonal reversal from south-westerly to north-easterly winds over the Arabian Sea. It now has much wider connotations.

In summer, land in the tropics becomes much hotter than the sea. Where the ITCZ is drawn into the summer hemisphere and pushes overland, increased uplift by hills together with surface heating generates some of the world's heaviest and most prolonged rain. The best-known monsoon affects the Indian subcontinent, where warm, moist air is drawn from over the Indian Ocean and the Bay of Bengal northwards across India, Pakistan and Bangladesh. This South West Monsoon breaks in late May or early June, giving months of heavy and often thundery rain. As the sun moves away southwards later in the year, the situation reverses. Air cooled over the Himalayan mountains and northern parts of the continent sinks, and is drawn southwards towards the now warmer sea. These North East Monsoon winds of late winter are dry and dusty in complete contrast to the summer rains. Monsoon conditions occur to a lesser extent in many other tropical regions, as the ITCZ wanders north and south of the equator. These include northern Australia, tropical Africa and South East Asia, but nowhere is the wet season as pronounced as the Indian subcontinent.

Rice is the staple food crop, and fast growing pulses and peanuts are important. Wild animals and vegetation in monsoon climates must be able to survive months of drought. Trees store water in large root systems, but most plant life becomes dormant, springing quickly into life with the onset of the rains. The savannahs of East Africa provide the habitat for some of the world's largest indigenous animals, where the occasional failure of monsoon rains provides a serious threat to their survival. In some cases, long-established migration patterns over many hundreds of kilometres enable them to thrive.

FURTHER READING

Meteorological Glossary, HMSO, London, 1991.
Comprehensive coverage of meteorological terms. Rather technical but excellent reference work.

The World Weather Guide, E. A. Pearce and C. G. Smith, Hutchinson, 1990.
Tables of temperature, rainfall and humidity through the year from places in all climatic zones around the globe, with covering notes.

Images in Weather Forecasting, M. J. Bader, G. S. Forbes, J. R. Grant, R. B. E. Lilley, A. J. Waters, Cambridge University Press, 1995.
Up-to-date and expert appraisal of satellite imagery interpretation in relation to all weather phenomena. Many excellent illustrations and diagrams. Expensive and technical, but invaluable for anybody wishing to get more out of satellite pictures.

Weather and its work, David Lambert and Ralph Hardy, Orbis Publishing Ltd, London, 1984.
A readable guide for the younger student through many aspects of global meteorology and geography.

Basic Meteorology – a physical outline, Robin McIlveen, Van Nostrand Reinhold (UK), 1986.
Detailed and comprehensive textbook, recommended for the serious and advanced student.

The Marine Observer's Handbook, 11th edition, HMSO, 1995.
More up to date than *The Observer's Handbook*, and preferred because of its excellent cloud photographs. Covers many aspects of practical weather observing.

Weather Lore, Richard Inwards, as updated by E. L. Hawke, SR Publishers, Wakefield, Yorkshire, 1969.
A comprehensive collection of sayings from the United Kingdom, America and Europe.

Handbook of Aviation Meteorology, 3rd edition, HMSO, 1994.
Very readable appraisal of meteorology as it impacts on aviation, but includes many aspects of wider interest.

Weather – monthly magazine of the Royal Meteorological Society.

Weatherwise – monthly magazine of the American Meteorological Society.

Both comprise varied articles and news items, usually non-technical.

Weather includes a summary of the previous month's weather over the United Kingdom highlighting items of interest, with daily thumbnail charts.

USEFUL INFORMATION

-------- **Useful addresses** --------

United Kingdom

The Royal Meteorological Society
104 Oxford Road
Reading
Berkshire RG1 7LJ

The society, founded in 1850, caters for enthusiastic amateurs as well as large numbers of professional weather men and women. It publishes professional journals and has many educational leaflets and pamphlets concerned with aspects of weather.

The UK Met Office Education Service
Johnson House
London Road
Bracknell
Berkshire RG12 2SY

Provides a wide range of teaching and reference material for schools and colleges, including videos and slides. Publishes leaflets and reports on major weather events and year-by-year statistics.

United States

The American Meteorological Society
45 Beacon Street
Boston MA
02108–3693

Serves the professional meteorological community of North America. Publishes meteorological journals and *Weatherwise* magazine.

The Internet

There are a large number of World Wide Web sites providing weather information, including reports, plotted charts, satellite pictures, forecast charts and written forecasts. More are being added all the time, so you will need to explore to find what you want. Most meteorological information is available through university sites, with by far the largest number in the United States. For North America you could start by browsing the 'Weather World' from the University of Illinois, which not only gives satellite imagery but movie sequences; also try the National Weather Service's Interactive Weather Information Network. In the United Kingdom, sample the Dundee University site for the most comprehensive library of satellite pictures, past and present, from orbits over and near the British Isles; the Universities of Edinburgh, Nottingham and Reading provide pictures from further afield.

Articles by R. Brugge in *Weather* magazine (see Further Reading) in November 1995 and April 1996 give the uniform resource locators (URL) of many meteorological Web sites.

INDEX